BRASSEY'S INFANTRY WEAPONS OF THE WARSAW PACT ARMIES

Infantry weapons, including infantry support vehicles, and combat aids in current use by the regular and reserve forces

Editor: Major-General John Owen, OBE, late Royal Marines

Compilers: Jimmy Barker, Sandy Cormack, Brian Edwards, John Marriott, John Simpson, Stanley Parr, John Weeks, Toby Wrigley

BRASSEY'S PUBLISHERS LIMITED
LONDON

Editor's note

This book *Brassey's Infantry Weapons of the Warsaw Pact Armies* is one of the three companion volumes designed to meet the needs of those who study the infantry weapons of the Warsaw Pact, NATO and Third World armies.

In response to demand, I have included a section on the workings of small arms and another on the mysteries of ballistics and hope that they will prove interesting and instructive. Also included is a penetrating account of the organisation, training and operating philosophy of the Warsaw Pact land forces by Lt-Col A.J. Barker. Its conclusions do not necessarily reflect the views of my publishers, however.

Information on Warsaw Pact forces is not easy to obtain and I am indebted, amongst others, to Captain D.A. Scott-Donelan of the Rhodesian Army for a number of excellent pictures and descriptions.

Designed by Eric Rose
Photoset by Moorgate Typesetting Company Limited
Printed and bound in Great Britain by
Morrison & Gibb Ltd, London and Edinburgh

BRASSEY'S PUBLISHERS LIMITED
10 Upper Berkeley Street
London W 1

First published 1976
Second edition 1979

Distributed by Croom Helm Ltd, 2–10 St Johns Road, London SW 11

ISBN 0–904609–35–9

Contents

Prologue

In NATO's early days the United States enjoyed a clear superiority in nuclear weapons and NATO strategy was based on the 'tripwire' concept, by which the primary function of ground forces in Europe was as a trigger for nuclear retaliation against a Warsaw Pact attack. However, as the Soviet Union has now reached nuclear parity with the United States, NATO's nuclear deterrent no longer carries the same dominant weight in the deterrent balance and this places a higher value on NATO's conventional military capabilities. Meantime technological advances in other fields, leading to the development of new weapons and new devices, have compelled changes in the organisation, structure and employment of conventional forces. Increased mobility, for example, allows combat elements to disperse over a wide front, so reducing the density of troops along the battle lines.

To be effective, any modern military force must be able to seize and maintain the initiative in the land battle. The ability to move troops, equipment and supplies on the ground through, over and beyond enemy lines is still necessary to achieve military success. In Europe, NATO and Warsaw Pact forces are organised and equipped for mobile warfare, and the numerical tank superiority of Warsaw Pact forces is unquestionable (7000 NATO tanks as compared with 21,000 deployed by the Warsaw Pact armies — a 3 to 1 margin). Whether this superiority in armour poses a serious threat is disputable. The record of the 1973 Arab-Israeli war shows that modern tanks are vulnerable to the non-nuclear, high technology, 'precision-guided' weapons which were then for the first time extensively employed on the battlefield. These weapons originated with such early devices as the French wire-guided anti-tank rocket of the 50s and the US 'smart' guided bombs used in the closing phases of the Vietnam War.

The apparent effectiveness of precision-guided weapons against massed tanks in the Middle East has led to intense development efforts; it also points to the value of the individual infantry soldier as a tank-killer. US forces in Europe already deploy Tow, Dragon, and Cobra/Tow devices; other NATO armies have deployed Milan. Several other more advanced NATO anti-tank weapons are in prototype form or undergoing evaluation tests. Another important anti-tank role is attributed to the guided artillery shell for 155 mm guns (code-named Copperhead). For their part the Warsaw Pact forces, equipped with Soviet anti-tank weapons, have a type of Panzerfaust — the RPG-7 grenade-launcher, together with 82 mm and 107 mm recoilless guns although these are being replaced by wire-guided Snapper, Swatter and Sagger missiles.

But it is unlikely that Warsaw Pact strategists visualize a simple conventional war in Europe where tank competes with tank and missile with missile. The disparity in tank strengths suggest that a Warsaw Pact blitzkrieg could overrun NATO forces in West Germany in a matter of days, unless it is stopped with tactical nuclear weapons. Realising this Soviet military doctrine asserts that Pact forces must be prepared for war to be nuclear from the very beginning of the conflict.

Where 'war in the main direction' is concerned, an official Soviet military writer suggests, 'one must regard the conduct of military operations as being the basic version'. And the fact that Pact forces are structured for nuclear warfare is consistent with this doctrine. The Warsaw Pact armoured and motorised divisions are organised on a Soviet establishment and are lighter than their NATO counterparts, comprising between 9,000 and 10,750 men. (As against 16,500 in a US division, 16,000 in a French division and 15,000 in one of the Bundeswehr.) In firepower and numbers of armoured vehicles, however, they are at least equal to NATO divisions. Exploiting a tactical nuclear strike, they could advance at an average speed of about 100 km per day on a 5 to 15 km front and covering a depth of between 10 and 15 km. Individual regiments are organised as battle groups, in size not very different from that of NATO brigades; like the NATO battle groups these regiments are capable of operating autonomously. Warsaw Pact tactical doctrine for 'dynamic' warfare calls for the employment of massed airborne or amphibious units together with huge armoured formations launching surprise attacks — perhaps with nuclear support.

Meantime enemy movement within the battle zone is halted by long range artillery and rocket fire and aircraft. The aim is to effect a rapid breakthrough and to destroy the enemy's tactical and operational reserves. Speed is the key, since it limits the enemy's ability to react and, above all, prevents him from using nuclear weapons against the advancing Pact units. An additional advantage, claim the Pact tacticians, is that a rapid sustained attack precludes an organised enemy withdrawal.

In the event of NATO forces responding to the attack with nuclear weapons, Warsaw Pact vehicles are expected to cross contaminated areas with hatches closed. For infantrymen who have to leave the protection of their APCs there is an adequate supply of nuclear, biological and chemical (NBC) equipment. And, if NATO forces deploy the controversial 'neutron bomb'— enhanced radiation warheads (ERWs)— as their prime anti-tank weapon, the Russians are likely to provide Pact armoured vehicles with added radiation protection. Furthermore, as exposure to neutron radiation would leave most of the armoured vehicles in a usable condition and the battlefield not heavily radio-active from fall-out, the attackers could be expected to have replacement crews ready to return the vehicles to battle quickly.

From this, it follows that the tank remains the hub of the war machines of both NATO and the Warsaw Pact. It has, in the words of Marshal Malinovsky, 'notable and multiple battle features which enable it to conduct any operation with success, even in a nuclear environment.' And, in common with all the other Soviet equipment supplied to the Pact armies, Soviet armoured vehicles are simple, inexpensive and efficient. In terms of armour protection their T-62 and the latest Soviet MBT T-72 tanks are superior to most Western tanks and they fall short of Western specifications in only a few other parameters. Thus the fact that they are deployed in overwhelmingly large numbers underlines the threat which NATO faces in Central Europe.

Another important consideration in assessing this threat which deserves mention is that of the rate at which opposing armies in Europe could be reinforced and resupplied. In the early weeks of a crisis the advantage of reinforcement and resupply would undoubtedly lie with the Warsaw Pact. This is largely because of the favourable geographical location of the Pact's dominant member, Russia, which keeps large reserve forces standing by. Much of the NATO equipment may be superior to that of the Pact and man for man the Western Alliance may be a better soldier. It can also be argued that an attacker needs a numerical superiority of 3 to 1 to be sure of success, and the Pact does not have this.

This then is the scenario against which the two alliances should be compared.

Introduction

The organisation of the Warsaw Pact bears superficial similarities to the NATO Alliance, but its military structure is very different. All key positions are held by Russians, and its headquarters are in Moscow operating as part of the Soviet Ministry of Defence, with East European officers serving mainly as liaison officers (such non-Russian officers normally receive extensive political indoctrination in the USSR before appointment). East European military resources, training and defence policy are coordinated from this headquarters. The Warsaw Pact, as such, has neither a command function nor a logistics establishment separate from that of the Soviet armed forces and there is no Warsaw Pact military doctrine distinguishable from that of the Soviet Union. The Russians are the dominant partner and Soviet military missions in East European capitals are believed to be responsible for the inspection and supervision of national training programmes and the standardisation of equipment and weapons. Thus the principal differences between the satellite armies and that of the Soviet Union stem from national and geographical characteristics; organisation, training, equipment and tactics are all based on the Soviet model (see table).

Form of service

All the member states of the Pact maintain strong defence forces and rely on conscription to provide the bulk of the manpower. Approximately 35 per cent of the total strength of the Pact armies is infantrymen — a slightly higher proportion than in the NATO forces. The length of national service varies; in the Soviet Union, Bulgaria, Czechoslovakia, Hungary and Poland a man usually serves for two years, while in the German Democratic Republic (GDR) he does 18 months, and in Romania only 16 months. Why Moscow should permit these deviations is not clear, but the reason may be a complex of factors relating to the national economies of Bulgaria and East Germany, the role of their armies in any future conflict, and their political reliability. The Bulgarian troops are usually tidily turned out but, by British standards, their uniforms are sloppy and their drill only fair in comparison with other Pact armies. Soldiers of the GDR on the other hand are equally smart, if not smarter, than their Russian counterparts — as anyone who has seen their infantry on ceremonial duties in East Berlin will vouch.

In all seven countries of the Pact every able-bodied male is liable for conscription on his 18th birthday. There are two regular call-ups every year in the spring and autumn, and it is mostly the poor and uneducated who serve. University students, and men who are the sole supporters of aged or invalid parents or have two or more children can have their call-up deferred. But university students can forfeit this privilege for bad behaviour or poor examination results. Pay is low, but food and living conditions generally equate to those of civilian life.

Russian soldiers get no leave during their national service and outside the Soviet Union they are virtually sealed-off from the population of the country in which they are serving. Troops of the other Pact armies are not subjected to such severe restrictions, mainly because they are stationed within the borders of their own countries and the need for social and political isolation is not seen to be so acute. In effect the communist conscripts have very little free time and, what there is, is closely supervised. Apart from political lectures, they are required to learn by rote the principal military regulations such as the rules for guard duty. Soldiers of all the Pact states are constantly reminded that the communist system is the best, and Russian troops are told that the Soviet Army is the best in the world — with the best equipment, best conditions of service, best food and so on. Most of the young soldiers believe this because they have not seen anything else. But those who are posted to units in Poland or East Germany are flabbergasted at what they see there.

Conscripts are initially assigned to the various arms on the basis of their educational qualifications, civilian skills, political reliability and physical characteristics — with the technical services having the pick of the crop. Men who have attended a secondary school are usually selected for NCO training, after which some are posted to infantry units. Drivers, of whom there is always a shortage in the Pact armies, are allotted to motorised or transport units. Politically unreliable conscripts are usually posted to infantry units in spite of skills that might normally qualify them for specialised assignment. As might be expected all the armies exploit the natural skills and backgrounds of certain ethnic or regional groups. In the Soviet Army, for example, Caucasians are usually sent to mountain units and Siberians are likely to serve as ski troops.

Promotion is governed by merit and, in the case of officer candidates, by the successful completion of a training course at an officer cadet school where — in addition to their military studies — they are instructed in the art of propaganda and political lecturing. More officers graduate from these schools than the seven Pact armies can absorb; because of this competition is keen. As an officer rises in rank, political reliability progressively becomes a more important factor in the promotion selection procedure and the great majority of officers of the Pact armies are Communist Party members or affiliates. Through the various staff colleges, the Soviet general staff at the Warsaw Pact headquarters is able to guide the selection and training of outstanding officers who are handpicked for duty with the general staff or earmarked for promotion to high rank.

Military training

Military training in the Warsaw Pact forces means considerably more than the mere acquisition of basic military skills and a knowledge of military regulations. To the political and military authorities in Moscow — to whom the Eastern bloc governments are all ultimately answerable — military training means the production of a *communist* soldier. And this is a process which demands the development of the correct ideological and political views, a high state of discipline, a capacity for self-sacrifice, unquestioning subordination to commanders and total loyalty to the Communist Party; complete dedication to the process of improving one's own performance and that of one's unit over and above the prescribed norms, unflinching courage in the face of danger and uncomplaining endurance during periods of deprivation and stress. All armies esteem most of these virtues, but only under the Soviet system are they so whole-heartedly pursued. Soviet doctrine stipulates that military training 'required at all levels' must never stop. There are, according to the pundits in Moscow, only two states of existence for a military man — either he is fighting or he is training. Clearly because of national and ethnic characteristics an ideal of this nature is pursued with different degrees of enthusiasm. The East Germans follow their orders with true Teutonic fervour; the Romanians are said to be least keen.

Training, carried on six full days a week is intense, continuous and realistic. Live ammunition is issued during field exercises and actual combat conditions are so closely simulated that training accidents are accepted as a matter of course. For the first year of an infantryman's service the Soviet Army's annual training cycle is divided into four phases. The first three months are devoted to physical conditioning, weapon training, range practice and marches carrying full equipment up to 25 km a day. The next phase, covering four months, starts with platoon and company training and concludes with battalion exercises. These continue into the third phase, which lasts another two months. Finally there are division and army manoeuvres to perfect all arms cooperation under conditions simulating actual warfare. Elaborate critiques are held at the conclusion of all major exercises and detailed correction is conducted by junior officers and NCOs.

The Soviet training programme is intended to produce efficient and tough soldiers integrated into well-organised cohesive battle teams. By the end of his first year the Pact infantryman is familiar with his weapons and knows how to use them to good advantage. He is conditioned to live ammunition and is inured to conditions as near to those on active service as his officers can make them. His physical condition has been toughened up and requires little of the elaborate care common in NATO's armies. If he has shown that he is a good shot, he may have specialised as a sniper; alternatively he may have learned how to handle one of the latest anti-tank weapons — the Snapper, Swatter or Sagger. On manoeuvres he rolls up his blanket (which is actually his greatcoat) under a bush, or joins his poncho cape to that of another soldier to form a bivouac tent. In the morning — if conditions permit — he has a hot breakfast (comprising a bowl of kasha, a chunk of black bread, and tea); otherwise his only hot meal of the day is a bowl of borsch in the evening. His peasant background generally permits an easy transition to field conditions and his stolid nature quickly adjusts to military routine.

Finally, mention must be made of the fact that Soviet operational and training directives all stress the theme of 'combat readiness' as an important objective in training. What this means was spelled out in 1972 by the Soviet Minister of Defence, Marshal Grechko: 'On any day or at any hour [combat-readiness] should be on a level that would make it possible immediately to begin and conduct the most decisive and effective actions in the event that the aggressors should start a war, and to change rapidly from one type of combat to another.'

Discipline and morale

Discipline, morale and military skills at both tactical and strategic levels are generally acknowledged as significant factors in assessing the potential effectiveness of armies. Soldiers of the Pact armies are subjected to a discipline of extraordinary strictness. Soviet soldiers have their heads shaved almost to the scalp during their conscript service — a measure which prevents desertion because a shaven head is an easily recognisable badge. Those who commit even the slightest violations of military discipline are severely punished. Signs of inebriation are punished with a minimum of ten days' detention; association with foreigners is a court-martial offence. And for court-martial offences, there is always the prospect of a disciplinary battalion — a labour camp almost identical to civilian labour camps. A deserter would be sent to such a place. But more simple punishments are known. If a cigarette butt is found on the ground, the soldier judged to be responsible may be ordered to carry it away on a stretcher to be buried in a trench dug by his entire company.

Disciplinary cases in the Soviet Army are sometimes transferred to the infantry. (Judging from a letter written by a Soviet major and published in the journal *Krasnaya zwezda* during September 1973 this is a practice strongly resented by the 'motorised rifle troops', as the Soviet infantry prefer to be called. 'We also strive to maintain high levels,' complained the major, 'and misfits from other services do not help.')

Morale among the rank and file varies. In the small GDR army, the infantryman is a relatively well-paid and dedicated communist who sees himself as one of the new élite; his morale is high. Polish private soldiers probably feel much the same, while Bulgarian and Romanian infantrymen have little enthusiasm for military matters. In the Soviet army the prevailing atmosphere in army life is usually boredom and there have been instances of men shooting themselves to get out of the army. (There is a simple rule for this: Fire at yourself below the belt and you can be tried for desertion; fire at yourself above the belt and it is considered a suicide attempt — for which you are demobilised and sent home.)

Among the NCOs of the Pact armies morale is usually good. The typical Soviet NCO is a hard communist-indoctrinated and determined individual who obeys orders promptly. His status is higher than that of most of his counterparts in NATO's armies, and considerable responsibility is delegated to him. As he progresses up the promotion ladder, he will attend a series of courses at one or other of the 'centres of instruction' which have been set up for training in the latest types of sophisticated weapons. For senior NCOs an obligatory nine months' infantry course includes the tactical employment of unit weapons and field engineering, as well as the inevitable political training. A great deal of time is also spent on the range. Junior NCOs attend less ambitious courses to learn about communications, battle drills and, again, how to be good Marxists.

One aspect of the communist control mechanism is the use of the soldier's family to restrain his behaviour and improve his performance as a soldier. Political officers often arrange to receive messages from a soldier's parents to inspire their sons to greater effort. It is characteristic of soldiers of some of the nations of the Pact — Poles and Czechs especially — to view this form of pressure with distaste. So far as the Soviet Union is concerned, under Soviet law the marriage and child maintenance allowances paid to a soldier's wife are stopped immediately if he is court-martialled and sentenced to serve in a disciplinary battalion. Treatment of this kind is counter-productive and there are indications that this is coming to be recognised.

Officers are in an entirely different category from the rank and file. In general the relationship between officers and other ranks in the Soviet Army is very formal and there is little contact. Clearly this state of affairs is not universally applicable throughout the Pact armies or even in the Soviet Army itself. But professional officers have their traditions and *esprit de corps* which are fostered by the communist regimes. Officers' social and professional contacts are mainly with their own kind and only occasionally will an officer drink with another rank in some private place.

Throughout his military career the Soviet officer is exposed to continuous political propaganda calculated to make him an enthusiastic supporter of communism. Even so he is constantly watched by his seniors, brother officers and the secret police for any departures from the approved Party line. There is an abundance of evidence to suggest that the same system is vigorously applied in the other Pact armies and is as effective — if not more so — in the armies of East Germany, Poland and Czechoslovakia as it is in the Soviet Army. In the past, the problems of political unreliability among officers and the reluctance of the Soviet regime to decentralise responsibilities have tended to stifle initiative. Current doctrine and training, as well as added perquisites of rank, are tending to increase the professional attributes of the officer class. In effect the training induces a double standard of compromising communism with military principle. It stresses flexibility in military matters, rigidity in political matters.

Despite these handicaps created by the communist system the disciplinary problems of the Pact forces are probably relatively insignificant. The limited free time available to both officers and other ranks, the tendency to impose group rather than individual activity and the large amount of time spent on manoeuvres and exercises and under close observation restrict the opportunities for committing violations other than those of a technical nature related to the soldier's duties.

Organisation

Up to and including the immediate post World War II era, the Russians always considered the infantry to be the main combatant arm; in the Red Army the infantry was the only arm which did not have a directorate in Moscow because infantry and Red Army were considered synonymous. This situation was partly a result of the great size of the infantry, and sheer weight of numbers ensured that most of the responsible command and staff appointments were held by infantry officers. Since 1944, however, the growing competition from the armoured forces undoubtedly caused the infantry to slip down in the league table of combatant arms. In the reconstituted satellite armies, except for that of East Germany, the infantry officer element has probably managed to retain more of its power than in Russia.

According to the latest information published by the International Institute of Strategic Studies (1978), the Warsaw Pact armies have a combined strength of 2,639,000 men of whom some 870,000 are infantry.* The operational units of this infantry are mostly deployed with motorised infantry divisions, armoured divisions and airborne divisions and, before considering the structure of these units, it should be noted that Pact formations are not all manned at the same level. They can be regarded as being in three categories:

Category 1 comprises units at full establishment strength with all their equipment;

Category 2 consists of units which are at between half and three-quarters strength, even though they may have all their equipment;

Category 3 comprises units at one-third strength — generally with all their equipment (although some of it may be obsolescent).

* These figures do not take account of Russia's 250,000 strong Marine Corps, of whom about half could be considered as infantry; nor of 200,000 KGB border guards and 250,000 MUD security troops. Poland also has a 1,000 strong force of Marines; Bulgaria, Czechoslovakia, East Germany, Hungary, Poland and Romania maintain sizeable paramilitary forces — border guards, security police, 'People's' or 'Workers' Militia', etc — all of whom are basically infantry.

Trained reserves which could be mobilised probably amount to about 6 million, of whom about 3 million will be infantrymen although in theory, because Warsaw Pact conscripts have a reserve liability up to the age of 50, the reserves could total 25 million.

The infantry elements of the motorised rifle divisions and the standard model Soviet armoured divisions are virtually identical. Infantry formations are organised on a triangular basis — three mechanised infantry battalions with an armoured division; three motorised infantry regiments (with three motorised battalions in the regiment) to a motorised infantry division. Supporting troops are proportionate to the role and importance of the division. The war establishment of an armoured division is about 9,000; that of a motorised infantry division about 10,700. The establishment of a rifle regiment is about 2,000 all ranks, and each of the three rifle companies in a nominal 500 strong battalion nominally contains 110 men. At full strength the company has three platoons of 30 men each of three nine-men sections; additionally every rifle company has a machine-gun platoon of 16 men divided into three sections with one machine gun each, usually the DShK 38/46.

Also included in the rifle battalion organisation is a heavy weapons company comprising a mortar platoon (with three 82 mm mortars, each served by five men), and an anti-tank platoon — the most modern of which now have the wire-guided Snapper, Swatter or Sagger missiles. All the men are carried in armoured personnel carriers (APCs) — armoured division battalions in tracked BMP-76 PBs, those in the motorised infantry divisions in wheeled BTR-60PKs.

The Soviet Union has eight airborne divisions, Poland has one, the Romanian Army maintains an airborne regiment, so too does Czechoslovakia. (The two latter are much the same size, but there are minor organisational differences). Airborne divisions — of which the Soviet Air Force can lift three simultaneously — have a strength of 7,200 men, and their combat element is based on three paratroop regiments, supported by an artillery regiment (of two battalions of three batteries with 85 mm or 122 mm guns and one rocket-launcher battalion of three 140 mm multiple rocket-launchers), together with an anti-tank regiment (one battalion of three batteries with 85 mm guns or wire-guided missiles). Each paratroop regiment comprises three battalions of 450 men, all of which have three companies. As Soviet paratroop battalions (and the reconnaissance battalions with the armoured and motorised infantry divisions) have been given the new BMD combat vehicle specially designed for airborne troops, the Russian airborne infantry will soon have increased mobility and firepower.* It is also worth noting that the Russians envisage employing one or two batteries with Frog tactical missiles with their airborne divisions — as with the Pact armoured and motorised divisions.

Finally the Soviet ski units, Romania's two mountain brigades and Poland's amphibious assault division deserve mention, since these are specialised infantry formations.

Weapons and equipment

The Soviet Union has been the arsenal of the Warsaw Pact since its inception in May 1945, and has either sold equipment to the satellite states or permitted them to build Soviet pattern equipment under licence. Besides benefiting the Soviet economy, this policy has achieved a high degree of standardisation of equipment and has virtually ensured that the satellite states are wholly dependent on the Soviet Union not only for initial supplies of weapons and equipment but also for spares.

The Russians have always tried to design simple, economical, robust and reliable weapons and equipment. Most of their designs are suited to mass production and where possible new equipment utilises proven sub-assemblies and spare parts available in equipment currently in service. One of the best examples of these endeavours is that of armoured vehicles. Thus the T-70 medium tank bears an obvious resemblance to the T-62 which also resembled its predecessor the T-55, and so on back to the grandfather of Soviet medium tanks, the T-34.

* The BMD, which appears to be derived from the BMP-76 PB, weighs about 10 tons and is amphibious. It has a 76 mm (or, according to some sources a 73 mm) smooth bore gun and a ramp for Sagger missiles; and it can carry six men in full equipment.

Soviet infantry progressed from riding on the hulls of tanks in the 1950s when the wheeled BTR-152 and the tracked BTR-50P were introduced. The former was little more than an armoured truck and the latter was just a turretless tank which could carry only seven men. Nevertheless both vehicles were milestones in the mechanisation programme of the Soviet Army, in the development of Warsaw Pact army equipment and also in infantry tactics. These vehicles have now been replaced in the Soviet Army, although some are still in service in the armies of other nations of the Pact. The Russian infantry — 'motorised rifle troops' — currently uses the 10 ton amphibious BTR-60PB (which can carry 14 men at a road speed of 72 km/hr (44 mph), at a speed in water of 9 km/hr (5-6 mph)) and the tracked BMP-76PP. The latter is one of the most advanced infantry combat vehicles in the world. Weighing just under 10 tons it can carry 8 men and is fitted with a 76 mm smooth-bore gun, Sagger wire-guided missiles and a heavy machine gun for AA protection against low flying aircraft and for use against enemy APCs.

The standard infantry weapon is the well-proven AK-47, the Kalashnikov assault rifle. Millions of these weapons have been manufactured and several variants of the basic design have been turned out in other Pact states. (In Czechoslovakia the variant is known as the M58 sub machine gun but it is identical with the AK-47. The Hungarian standard 7.62 mm sub machine gun is also basically an AK-47 although a hollow plastic butt makes it lighter than the Soviet version.) The original Kalashnikovs were tested and tried during World War II, and when one fell into the hands of the Germans they were so taken with the design that they modelled their Sturmgewehr on it.

Warsaw Pact sharpshooters are armed with the SKD (*Samosarjadvy Karabin Dragunov*) or SVD (*Snayperska Vintovka Dragunov*) Dragunov sniper rifle, an outmoded bolt-action rifle fitted with a telescopic sight and firing the old 7.62 mm rimmed cartridge used in the World War II (and earlier). Mosin-Nagant rifles are still issued to some of the Pact's para-military forces. In the Soviet Army Kalashnikovs have totally replaced the PPSh (*Pistolet-Pulemet Shpagin*) and PPS (*Pistole-Pulemet Sudarev*) sub machine guns of World War II but both the PPSh and the PPS continue to be used by the armies of Bulgaria and Hungary (the latter manufactures a local version of the PPSh). Para-military and reserve troops of all the Pact states are also likely to have considerable stocks of these old weapons.

A variety of pistols — personal weapons issued to officers and administrative troops — exists. The more common Soviet weapons, some of which have been taken from terrorists (freedom-fighters) in southern Africa, are the 9 mm Makarov and the 9 mm APS (*Automatitsheski Pistolet Stetshkin*); the Czechs, the East Germans and the Poles also produce locally designed pistols as well as copies of the Makarov and APS.

Machine guns differ according to the particular army, but Soviet designs predominate. In the light machine gun range the Soviet RPK (*Roshnoi Pulemet Kalashnikova*) — an elaboration of the AK-47 assault-rifle, the RPD (*Roshnoi Pulemet Degtyarev*) — an obsolescent weapon in the Soviet Army deserves mention. The Czechs also possessed the locally designed and manufactured M1952/57 LMG.

In the medium to heavy machine gun class the 7.62 mm Soviet SGM (*Stankovy Goryunov Modernizovannyi*), the Soviet 12.7 mm DShK (*Degtyarev-Shpagin*) and the 14.5 mm KPV (Vladimirov) weapons are the most important. The latter two heavy guns, the DShK and the KPV, are mounted on APCs and used only by the infantry for AA protection and against enemy APCs. Apart from their M52 and M57 LMGs the Czechs also manufacture a heavy machine gun — the Kulomet ZB-60.

Infantry mortars in service are the 82 mm M-42, the 120 mm M-43. The former, and its predecessor the M-37, is widely used by infantry of the Pact armies other than that of the Soviet Union where it is largely confined to paratroop battalions. (Other Soviet infantry units and infantry of the East German Army employ the 120 mm weapon.)

In the anti-tank field there are the Soviet 40 mm RPG-7 grenade launcher (*Reaktivny Protivotankovyi Granatomet*) — a bazooka type weapon which can penetrate 250 mm of armour at a distance of 300 m — the recoiless 82 mm B-10 and the 107 mm B-11 recoilless anti-tank guns. But all three types of weapon have largely been replaced by the wire-guided Snapper, Swatter and Sagger missiles in all Soviet regiments (other than air-borne regiments).* The other Pact armies are following suit but re-equipping will take some time and so the obsolescent RPG-72, B-10 and B-11 guns are likely to remain in service for some considerable time. Czechoslovakia produces its own anti-tank rocket projector (the P-27 *Pancerovka*) and 82 mm light anti-tank gun (the T-21 *Tarasnice*) as could be expected of a country which has an old and well-established armaments industry.

A recent addition to the Warsaw Pact infantryman's arsenal is the SA7 Strela, a man-portable, shoulder fired surface-to-air missile which homes on to aircraft exhausts. The record of the October 1973 War in the Middle East suggests that this weapon is particularly effective when an attacking plane is at an altitude when electronic countermeasure devices are not being used.

Information concerning technological trends in the Warsaw Pact countries is difficult to come by. The development of a new weapon or new equipment is rarely mentioned in Soviet literature or by the press, and often the first evidence of its existence is its appearance on Moscow's Red Square in a May Day parade. But Moscow takes a keen interest in Western technology and the performance of such weapons as the AR15 (M16) rifle used by the US in the later stages of the Vietnam War has not escaped them. It must also be remembered that the Soviet Union obviously gained considerable practical knowledge relative to the performance of the vast array of its military equipment used by the Arab forces during the October War in 1973. In Africa the East Germans and the Soviet Union's Cuban surrogates are accumulating further knowledge that will eventually be used to improve weapons in service, to develop specifications for new equipment and to formulate new tactical doctrines.

Tactics

Warsaw Pact infantry training is carried out against a background of nuclear and chemical war, and is designed to cater for both nuclear and non-nuclear warfare. The tactics employed are concerned primarily with the offence and are based on Soviet doctrine of a war of movement. Basically this resolves into an initial strike in great depth and a rapid advance by mobile armoured forces deep into the enemy rear. Moving on independent axes, accepting the risk of open flanks and complete separation from the main body, the aim of such formations is to cover about 100 km (60 miles) a day. Night fighting and movement by night are much favoured although continuous round-the-clock fighting by all arms combat teams is anticipated. In a deliberate attack, tanks and APC-borne infantry are trained to make a rapid advance and deploy in waves — being covered by the fire of artillery, stationary tanks and SP guns. Infantry may use their APCs as battle taxis, dismounting short of the objective and making the final assault on foot. Judging by the performance of the Soviet-trained Syrian infantry in the October 1973 Arab-Israeli War, however, it seems more likely that they may motor onto or through the objective. If resistance is light, APC-borne infantry can be expected to shoot from their APCs and dismount only when they reach their objectives. Minor tactics are essentially offensive in character, based where possible on outflanking or encircling movement.

Personal equipment and mobility

In World War II the Red Army infantrymen marched more miles than any of his allies. He was able to do this because he carried only basic equipment, minimum rations and very few of the luxuries which his western counterpart tends to rate essential. Today the Warsaw Pact infantryman does not expect to do so much marching as his predecessor, but he trains with a pack, which may weigh as much as 35 kg. But the Warsaw Pact High Command appears to envisage its infantry fighting light, and personal equipment and battledress appear to have been designed to meet the needs of soldiers whose mobility will not be impaired by an overstuffed pack. (It is relevant to add that on past experience, however, the Russian fighting man does not readily abandon equipment; on the contrary, he is more inclined to scavenge clothing, food and items to which a NATO infantryman would not give a second thought.)

* Snapper, Swatter and Sagger are the NATO codenames. Snapper is known to the Russians as *Schmel* (Humming Bee); the other two are merely given service designations.

Battledress is as simple and practical as the Soviet and East German Armies' dress uniforms are smart and colourful. Being full and loose fitting it is well suited to crawling and creeping around difficult terrain. High leather boots are normal issue because they have proved their worth in Central Europe and on the Russian steppes where the mud is knee-deep in winter. For soldiers operating in snow-covered regions, a loose-fitting white cape is issued and fur boots (*valenki*) are worn in place of the normal leather issue in dry sub-zero zones. Minor differences exist between armies: for example the East German infantryman wears socks, while the Soviet soldier wraps foot cloths, called *portyankai*, round his feet, although this practice is going out of fashion and Russian soldiers are now being issued with socks.

Personal equipment includes grenade and ration pouches, a crude but practical rucksack, a mess-tin, water bottle, poncho (rain cape) and a long skirted greatcoat. The last, serving as a coat and blanket, is regarded as the most valuable piece of the Pact infantryman's equipment — next to his rifle and water bottle. It is carried — slung in a horse-shoe roll over the shoulder — even in the heat of summer and in World War II, the Germans found this habit to be useful in distinguishing enemy soldiers from their own at long range. Two poncho capes are designed to fit together to form a tent, but with his genius of improvisation the Soviet soldier has shown that he can usually improvise bivouacs and shelters from material 'liberated' (even in peace-time Europe). No doubt his skills in this respect have been communicated to other Pact infantrymen.

The 4.2 kg rugged and reliable Kalashnikov rifle (*Automat Kalashnikova*) constitutes the main item of a combat kit which weighs about 25 kg in summer and 32 kg in winter. The rest comprises 120 rounds of ammunition (in four magazines) two grenades, a full water bottle, field dressing, entrenching tool, rations and mess kit. With the greatcoat this gives a total of about 22 kg; toilet articles, tobacco and extra clothing account for the other 3 kg. Extra clothing is not always carried, however, and the rations on manoeuvres may be reduced to a piece of bread. Full personal NBC and radiac equipment has been carried in Pact APCs in recent years — giving the Pact infantryman an extra capability for defence against NBC attacks, or perhaps for exploitation in the event of the Pact High Command deciding to employ NBC warfare.

According to Soviet doctrine paratroops jump with a heavy load — but fight light. The average Soviet parachutist jumps with a 27 kg load, exclusive of his parachute; the lightest equipped are officers and senior NCOs who jump with a load of about 24 kg.

Finally rations — for these are a factor equally as important to the efficiency of the Pact infantry as to other armies, and their form is necessarily conditioned by taste. As this is subject to regional variation, it is to be expected that the rations issued to men of the East German Army, for example, may differ in type and quality — and indeed in quantity — from, say, those issued to Hungarians. But there is one overriding consideration common to all field supplies issued to Pact armies; consistent with sustenance, they are simple and frugal. The Soviet infantryman going on manoeuvres is issued with a quantity of kasha and bread sufficient to see him through the duration of the exercise. Re-supply is a secondary consideration and the soldier is expected to fend for himself — foraging for fuel to cook his meal and anything extra he needs. This he usually does by stripping orchards and farms in the exercise area and hacking down fences for fuel.

Summary and conclusions

The Warsaw Pact armies are based on the Soviet model, and military doctrine is also dispensed under monopoly conditions. Soviet propagandists emphasise the importance of the Pact and the common military effort of the socialist bloc. 'Military collaboration,' they say, 'provides a sense of common purpose and feeling.' To ensure that this sense of common purpose will continue, however, Moscow plants her own nationals in key posts in the satellite armies and, although several of them have now developed their own trappings, such as uniforms, everything else is very heavily under Russian influence.

The common armament policy has some exceptions — notably in the field of small arms, radio equipment and soft-skinned and support vehicles. (Czechoslovakia has produced an AFV — the amphibious APC OT-64, which is also used by the Polish army and known there as the Skot. Hungary also produces its own armoured amphibious scout car.) But the centre of gravity of the Pact's military strength is firmly fixed in Moscow.

On the whole, equipment and organisations are excellent, morale is generally high, training is realistic and the troops tough and accustomed to hardship and harsh conditions. There is no evidence to suggest that their shooting is any better than that of NATO troops, but the Pact armies undoubtedly train more industriously than their easier-going Western counterparts. The fact that the Soviet Pact armies have not seen any fighting since 1945 while some of the NATO armies — notably those of Britain and France — have been involved in military operations, albeit 'colonial-type' wars, may also be relevant. Limitations attributable to the communist political system and the personnel selection system can also be expected to reflect in unimaginative tactics and leadership at platoon and section level. Thanks to hard training plus conscription and large armies in being, however, the Warsaw Pact has a tremendous pool of trained manpower. So the Soviet High Command is in a position — and possibly sees a need — to expend its men lavishly, even in the nuclear era.

Warsaw Pact Army strengths

COUNTRY	ARMY STRENGTH	RESERVES	OPERATIONAL FORMATIONS	INFANTRY (ESTIMATED)	MILITARY SERVICE
Bulgaria	115,00	200,000	8 mot rifle divs	36,000	2 years' conscription
			5 tank bdes	2,500	
			1 airborne regt	1,400	
Czechoslovakia	140,000	300,000	5 tank divs	25,000	2 years' conscription
			5 mot rifle divs	2,500	
			1 airborne regt	1,400	
Germany (Democratic Republic)	105,000	250,000	2 tank divs	25,000	18 months' conscription
			4 mot rifle divs	20,000	
Hungary	91,000	130,000	1 tank div	2,500	2 years' conscription (incl border guard duties)
			5 mot rifle divs	23,000	
Poland	222,000	500,000	5 tank divs	2,500	2 years' conscription
			8 mot rifle divs	40,000	
			1 airborne div	4,100	
			1 amph aslt div	3,000	
Romania	140,000	300,000	2 tank divs	5,000	16 months' conscription
			8 mot rifle divs	40,000	
			2 mountain bdes	3,000	
			1 airborne regt	1,300	
USSR	1,825,000	*	46 tank divs	95,000	2 years' conscription
			115 mot rifle divs	500,000	
			8 airborne divs	35,000	

*Reserve obligations for Army, Navy and Air Force to age 50 mean total reserves could be 25,000,000.

Summary	Total Warsaw Pact numerical strength :	2,639,400
	Infantry strength :	870,200
	Percentage infantry :	33%

Mechanisation and technological development since World War I is reflected in the proportion of infantry then and now. In 1917 approximately 70 per cent of the 15.5 million men mobilised by Imperial Russia were infantry.

Ballistics Simplified

The dictionary defines ballistics as 'the scientific study of the motion of projectiles' and gives the root as the Greek word *ballein*: to throw. Ballistics is therefore the study of what it is that makes a projectile, a bullet, a mortar bomb, or a rocket, actually go in the way that it does. It is a large subject which can become incredibly complicated and obscure if it is followed to its logical conclusions. Fortunately for this broad survey, it is both possible and sufficient to keep to the simplest outline explanation.

The first point about ballistics is that it is a study that is divided into two parts, interior ballistics and exterior ballistics. In other words, what goes on inside the gun and what goes on outside the gun, once the bullet has left.

In this very general account of ballistics it will be convenient to use small arms as the example, but in broad terms the same actions occur inside large guns also, though, of course, the times and forces involved are changed. Low velocity weapons such as mortars are very similar in their ballistics but, where there are differences, these will be brought out. But the general example will be a modern rifle of 7·62 mm calibre, firing NATO ammunition.

Interior ballistics

Interior ballistics is the study of the events which occur in the crowded instant between the release of the striker or hammer by the sear and the ejection of the bullet from the muzzle. The various happenings, taken one by one, would be quite simple to understand were it not for the incredibly short space of time in which they happen. Since time is the essence of this whole aspect of the bullet's flight, it is as well to begin by getting some idea of just how brief it is. From the release of the striker to the sound of the shot reaching the firer's ear is no more than one hundredth of a second. To the actual instant when the bullet leaves the muzzle it is slightly less, about three-quarters of a hundredth. So much happens in that short time that it is easier to divide the hundredth up into ten parts of a thousandth of a second each. We can then say that it takes about eight thousandths for the bullet to leave the muzzle, and a further two thousandths for the sound to reach the firer.

It is now convenient to divide the actions which are covered by interior ballistics into four periods, and to describe each in more detail.

The first period is the striking of the cap in which the hammer or firing pin comes forward and hits the cap a sharp blow. The actual time taken for this is usually referred to as 'lock time', a short lock time being ideal as it gives the least opportunity for the firer to flinch and affect his aim; as a general rule, lock time on modern weapons is around five thousandths of a second. The striker has to give the cap a sufficiently hard blow to crush the powder composition and cause it to detonate from the friction. The same thing happens on a smaller scale with a child's cap gun. The force of the blow has to be controlled to within certain limits; too much will pierce the cap and allow gas to blow back, too little can cause partial detonation of the composition and slow down the firing of the main propellant.

The second period is the explosion of the cap composition and the ignition of the propellant. Caps are no longer filled with mercury fulminate, they now use modern mixtures such as azides which have a better storage life and which do not leave corrosive fragments in the barrel; but the effects are the same. On firing, the composition turns into white hot flame, as hot as an electric welding flame, and of sufficient volume to raise the pressure inside the cartridge case to about a ton to the square inch. This intensely hot flame spreads throughout the case and envelops every piece of propellant, igniting it immediately. It makes no difference whether the propellant is a powder or in sticks, it is all evenly lit at the same instant. Once lit, it burns at an increasing speed. The slow start is due to the fact that the gas has to warm up the brass case and the unburnt propellant, and so it loses some of its energy in doing this. Once all is warm, the burning becomes so fast that it is pretty violent and the pressure quickly builds up sufficiently to move the bullet forward out of the neck of the case. This second period of time is one which is very difficult to examine since all the action is confined inside the case, but it can be estimated with some accuracy that it all happens within about half a thousandth of a second.

The third period is that during which the bullet is moving up the barrel. It starts just as the bullet begins to move. This usually needs about a couple of tons/sq in to get the movement started, but the pressure inside the case may rise to as much as 5 tons/sq in before the actual movement becomes apparent and the inertia of the bullet is overcome. In that very short space of time, the base of the bullet will expand slightly and make a fully gas-tight seal in front of the gas. If it does not expand, gas can leak past it and go up the barrel in front of the bullet, which will cause it to seat unevenly in the rifling and give alterations in the muzzle velocity and the alignment along the trajectory. As the bullet moves, the propellant continues to burn and make more gas until a point is reached where the pressure is at its greatest, about 22 tons/sq in with NATO 7·62 mm cartridges. From then on, the bullet moves faster than gas is being made and, shortly after peak pressure, all the propellant is burnt and the pressure steadily drops until at the muzzle it is about 6 tons/sq in. The bullet continues to accelerate because there is only friction for it to overcome and it leaves the muzzle at 1,800 m/sec, or thereabouts, there being slight differences in speed for different barrel lengths.

The fourth and last period is the brief interval when the bullet leaves the muzzle and flies through the disturbed air and gas which ejects behind it and catches it up. The remaining gas in the barrel takes a further three or four thousandths of a second to come out of the muzzle, and then the entire operation is over. The sound reaches the ear of the firer and, at the most, one hundredth of a second has elapsed since the striker slipped off the sear.

Modern smokeless propellants contain enormous amounts of energy. The propellant in a NATO 7·62 mm case is capable of producing roughly 20 horse-power. Of this, only one-third is used to propel the bullet; slightly under one-third goes as heat into the barrel; and slightly over one-third goes out of the muzzle as blast, flash and noise. With a machine-gun firing 600 rounds a minute, the horse-power going into the gun is about 200, and 60 horse-power of heat is going into the barrel!

For big guns the interior ballistics are very similar, though the time intervals are longer and the energies are greater. By and large the pressures are the same, but there is no expansion of the base of the shell. In a mortar the peak pressure is usually in the region of two to four tons, much the same as in shotguns, and the projectile is much slower, almost always below the speed of sound. Otherwise the same rules apply.

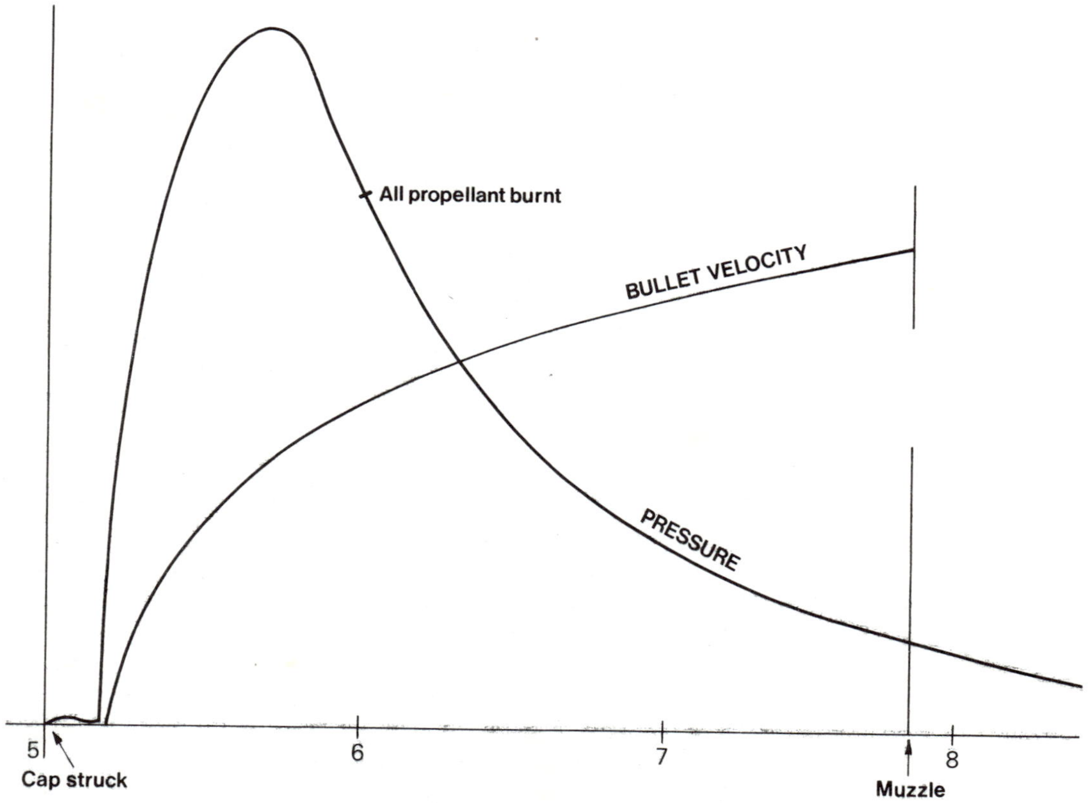

A diagram of the happenings inside a gun on firing. Time along the lower line is in thousandths of a second, and the diagram starts at 5 thousandths, when the cap is struck. There is then a very short pause and the pressure rises sharply as the propellant burns rapidly. The peak pressure in this case is over 20 tons/sq in. Shortly after that all propellant is burnt and the pressure drops steadily. The bullet starts to move shortly after the pressure rise starts and accelerates steadily to the muzzle

Exterior ballistics

Once the bullet leaves the muzzle the laws of exterior ballistics take over. These can be studied with much greater precision because the flight can be observed and, with modern high-speed photography, any bullet or shell can be watched as it leaves the barrel—or even followed down the range. For all general cases, the flight of a bullet is over in less than half a minute but this offers a very different time frame from that applying to interior ballistics and it allows plenty of time in which to decide what is happening.

The curving flight of a bullet is called the trajectory, from the Latin words *trans* (across) and *jacere* (to throw). A true trajectory, which only occurs in a vacuum where there is no air resistance, is a geometrical curve known as a parabola; but an actual trajectory is a modified parabola in which the latter part of it curves more steeply downwards due to the bullet slowing down and dropping. In a vacuum, maximum range is achieved by firing at an angle of 45°, but in air this does not work and the angle becomes 33°, a rough rule which holds good for most guns. As a crude approximation, the angle of descent is $2\frac{1}{2}$ times greater than the angle of launch, so for 33° elevation of fire, the bullet will arrive at 82·5°, or very nearly vertical. Any greater elevation of the gun merely means that the bullet will actually drop vertically and the last part of the flight will add nothing to the range.

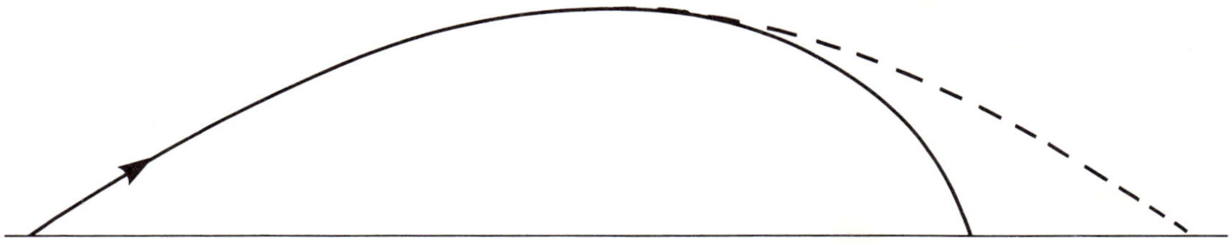

The dotted line is a trajectory in a vacuum. The solid line shows how it is modified by air drag

The early flight of any bullet after leaving the barrel is unstable. In that period it is recovering from the shock of discharge and the upsetting influences of the muzzle blast, so it generally goes off down the range with a slight yaw or 'wobble' which straightens out after about 100 m, whereupon the bullet settles down to the main part of its flight and flies very much as its designer intended it to; that is to say nose first, spinning steadily and with its centre line running along the trajectory. During this part of the flight it is possible to calculate the behaviour of the bullet with a good deal of accuracy and to check these predictions by experiment. This second part of the flight is the useful part of the bullet's life, and the one which the firer endeavours to use to the full. The last part of the trajectory is the final slowing down and drop out of the sky, which has been mentioned already. This is brought about by two things, air drag and yaw.

Air drag is the natural resistance of the air to a moving body and, at the high speeds at which bullets fly, it is sufficient to bring about a steady loss of velocity to the point where the path of the bullet curves downward more and more steeply. A second influence is that of yaw. The spin of the bullet has to be arranged so that it is greater than actually needed in order to make sure that there is enough stability to get through the first part of the flight and without being affected and turned off course. In the last part of the trajectory the bullet is usually too stable and tries to remain facing in the direction that it has done for most of the trajectory. As the trajectory steepens downwards the bullet begins to fall almost sideways on, which builds up more drag and can lead to it tumbling base over apex. This of course only happens at extreme range where few firers are trying to shoot anyway.

The spin given to bullets is considerable and the NATO bullet leaves the muzzle rotating at about 3000 revolutions per second and it loses little of this spin during the flight. Big guns are very similar, though they have to take care how they spin their shells in the barrel since it is quite easy to rip off a driving band with too much twist. Rifle bullets, on the other hand, can stand quite sharp twist angles as they grip the rifling by deforming the outer skin. Mortars, with the notable exception of the few rifled mortars in existence, have no rifling and rely on fins for their stability. These give a much greater upsetting motion at the muzzle as the blast overtakes the bomb in the first couple of metres of flight and causes it to fly backwards in relation to its surrounding air. This causes the well-known wobbling flight of mortar bombs, which the fins have to damp out as quickly as possible before too much drag builds up. Indeed, the designer has to compromise between large fins, which cause drag and muzzle upset, and small fins which do not, but which fail to smooth out the flight before drastic yawing sets in.

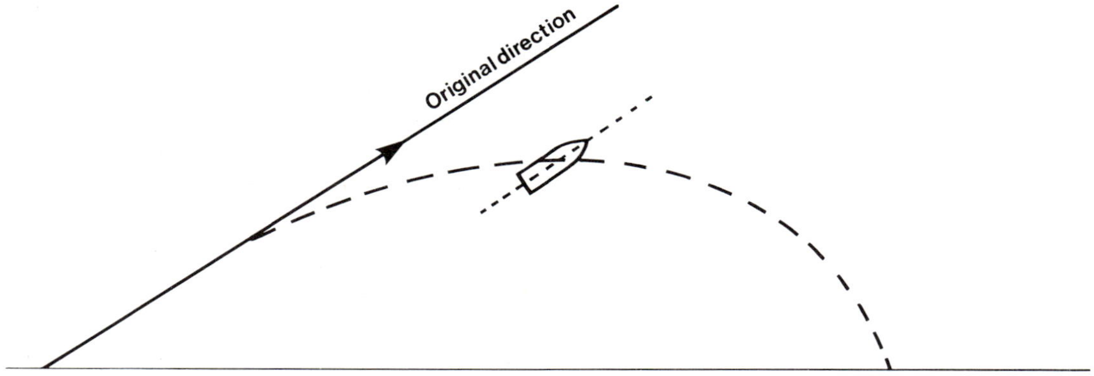

An exaggerated view of an overstable bullet, showing how it can fly almost broadside on at the end of the trajectory

Bullets are made in the shape they are to reduce drag and improve their penetration. The ideal bullet shape is an arrow. It has a small frontal area and so very little drag, but there is a great deal of weight behind the small front. If rifles could be made to fire metal arrows it would be ideal from the point of view of exterior ballistics, but there are severe practical difficulties in getting a long projectile to fly in a stable fashion without having to add drag in the form of fins and unless great care is taken all the advantages are quickly lost, quite apart from the virtual impossibility of making a rifle of suitable size and weight to shoot a long projectile. So arrows, though theoretically perfect, are practically impossible and the present pointed short cylinder is the best compromise that can be found to give all the desired results. It has the great advantage of being cheap, and that counts for a lot when one considers that armies use small arms ammunition by the millions of rounds every year.

The Operation of Small Arms

Ammunition

Small arms ammunition in use at the present time is all of the conventional type. This is to say that it all has a case, a primer, a propellant and a projectile.

The cartridge case is normally constructed of either brass or steel, except in special purpose rounds such as blank or practice, where plastic is often used. The case performs functions other than as a simple container for the components of the round. It must be strong enough to withstand high pressures, yet be flexible enough to seal the breech by expanding against the chamber walls. The case also protects the chamber walls from the direct action of the burning propellant and on extraction removes up to 30% of the heat from the weapon.

The primer in small arms ammunition is of the percussion type. It contains a sensitive explosive compound which in the case of the Boxer primer is sandwiched between the primer body and a built-in anvil. In the Berdan primer the anvil is part of the cartridge case. The action of both types is the same in that the striker or firing pin detonates the primer by driving against the primer body and forcing the compound against the anvil.

The propellant is an explosive whose burning rate is slow enough to be contained in a weapon without the pressure reaching unmanageable levels. See 'Ballistics Simplified' on pages 1–3.

The projectile has a construction that varies according to the desired terminal effect. The standard round is ball; if an indicator of the direction of the fire is wanted, tracer is used. Hard targets need an armour piercing capability and some soft targets require incendiary. Occasionally in heavy machine guns an explosive round is also available. The round can have a combination of characteristics such as armour piercing/tracer or explosive/incendiary.

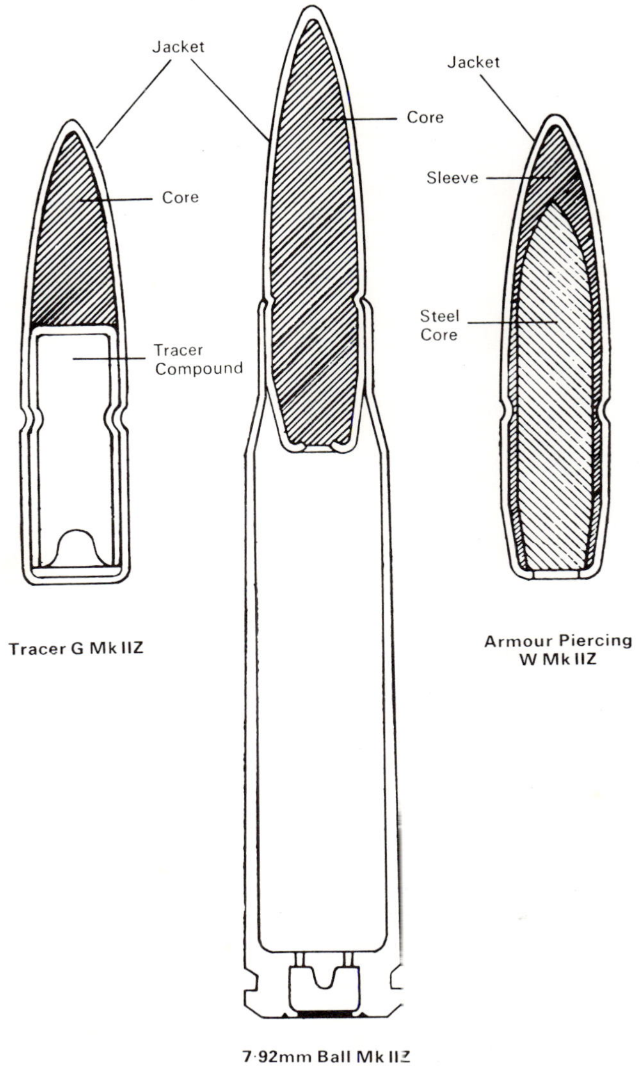

Any weapon whether operated by hand or automatically has the following simplified cycle: load - fire - eject - reload.

Single shot

The M79 grenade launcher illustrates the simplest form of weapon. It is a hand loaded single shot with a standing breech and a barrel that hinges downwards for loading. The empty case is ejected on opening.

The revolver

The modern revolver has a standing breech with the cartridges contained in a cylinder that is rotated automatically by the action of the trigger. The solid frame revolver, such as the Colt or Smith & Wesson, has a crane on which the cylinder may be swung out sideways for loading. The break frame type such as the Webley, tips downwards on a pivot for loading. The revolver may be fired single action ie the hammer is cocked by hand, or double action when the hammer is cocked by the action of the trigger.

The bolt action

There are three basic types of bolt action. First, the front locking turn bolt type, exemplified by the Mauser action in which the locking lugs on the bolt lock directly into the barrel extension. Secondly, the rear locking turn bolt type, such as the British L39A1 rifle in which the locking lugs are in the main receiver body. The third type, which is currently only in use by the Swiss Schmidt Rubens M31/55 sniper rifle, is the straight pull bolt. On this type, as the name suggests, the bolt is pulled to the rear and pushed forward without any turning action.

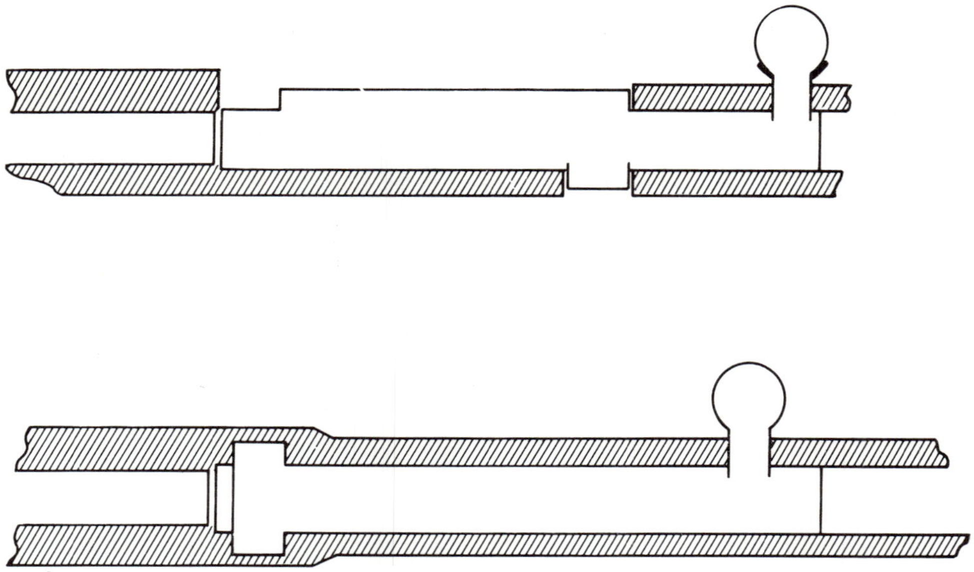

The top diagram shows the rear locking turn bolt type where the bolt lugs engage into the rear of the receiver. The British No 4 rifle is a typical example
The lower diagram is of the Mauser front locking system where the bolt lugs lock directly into the barrel extension

The logical development in weapon design was to use either the recoil energy or the gas pressure to operate the weapon cycle. The type of action can be divided into three groups: blowback, recoil and gas. All these groups have sub-divisions.

Simple blowback

The simple blowback utilises the gas pressure to drive the cartridge case and thus the bolt backwards. Only the inertia of the bolt and the recoil spring hold the breech closed. The rearward movement of the bolt ceases when the recoil spring overcomes its momentum. An ejector throws the case clear during this rearward movement. The bolt driven forward by the recoil spring removes a round from the magazine and chambers it.

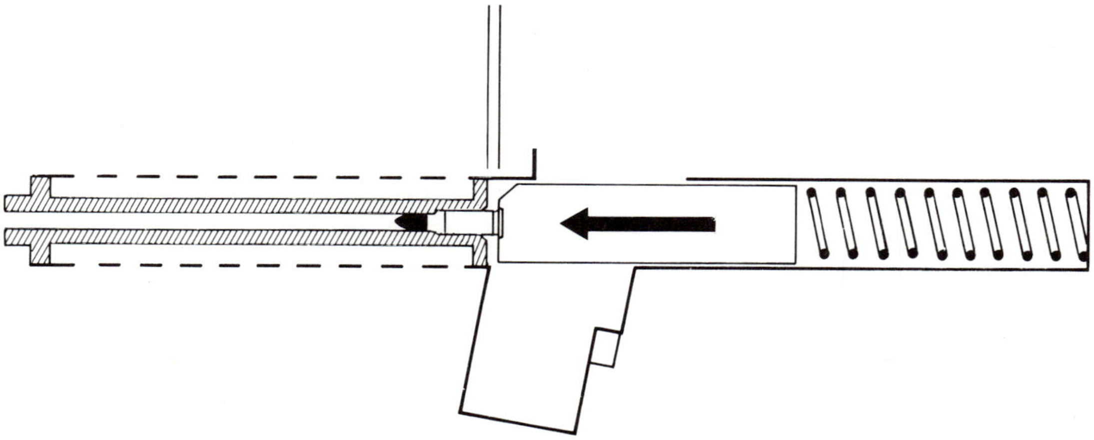

Advanced primer ignition is typified by the Sterling sub machine gun. The bolt is still travelling forward when the cartridge is fired

Blowback with advanced primer ignition

The bolt, having the energy of the recoil spring driving it forward, fires the cartridge while it still has .08 of forward movement. This means that the gas pressure now has to overcome the momentum of the bolt, then its inertia when it is stationary, along with the recoil spring force before it can move rearwards. The extra energy required allows a lighter bolt to be used. This is typified by the Sterling submachine gun.

Retarded or delayed blowback

Straight blowback and advanced primer ignition are only suitable for low pressure cartridges and thus are used only for pistols or submachine guns. To enable the pressure of a high pressure cartridge to drop to a safe level for blowback to commence, some form of delay must be used.

Roller delay

A typical example, the HK G3 rifle, utilises a two piece bolt, the front half of which holds the two rollers. When the bolt is forward with a round chambered, the rear component of the bolt driven by the recoil spring forces the rollers outward into recesses in the gun body. On firing, the front component tries to move rearward but to do so it must force the rollers out of their recesses. The movement has in turn to force the rear component back against the recoil spring. The time taken to achieve this is sufficient for the pressure to drop to an acceptable level for blowback.

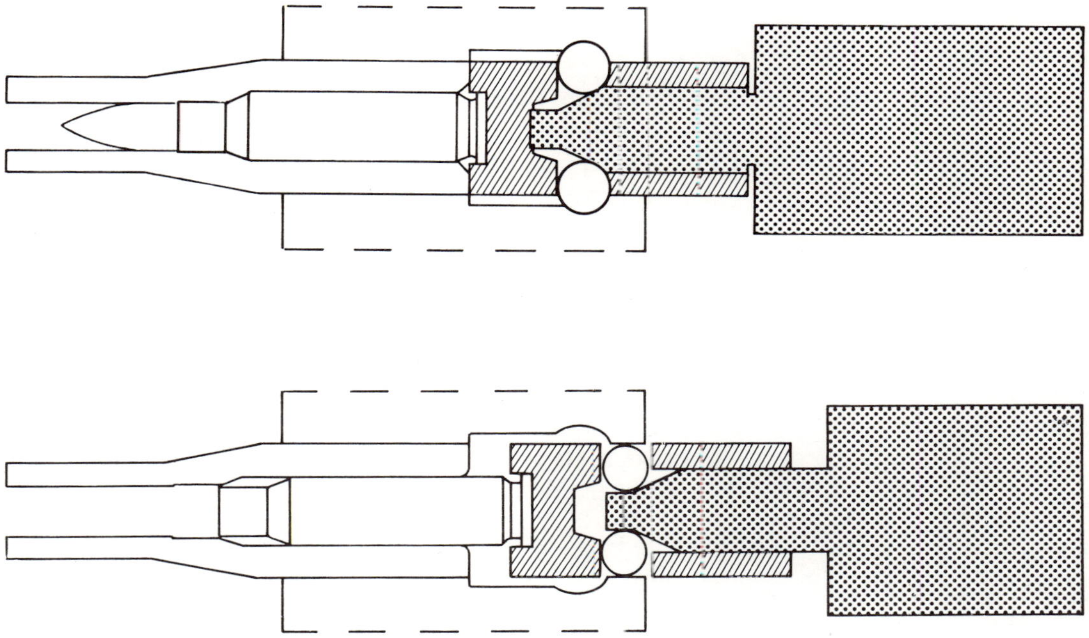

Roller delayed blowback is a feature of the Heckler & Koch weapon systems. The top diagram shows the weapon ready to fire with the rollers pushed by the rear component of the bolt into the recesses of the barrel extension. The lower diagram shows the weapon fired and the roller being forced out of engagement by the front component of the bolt

Lever delay

The basic operation of the lever delay is exactly the same as rollers, but substituting either a single lever such as in the MAS 5.56 or compound levers such as in the SIG.510.

Fluted chambers

A feature of many delayed blowback weapons is the use of a fluted chamber. The flutes which run from the front to the rear enable gas to bleed back and float the case off the chamber walls. This eases the problem of initial extraction which could cause case head separation.

Recoil operation

The use of recoil operation is largely confined to pistols and machine guns. Recoil operation divides into two major types, long and short.

Long recoil: This features the breech remaining locked on the rearward movement of the barrel and bolt group. The unlocking occurs with the group at the rear and ejection occurs with the forward movement of the barrel. An example of this type of operation is the Rarden cannon where it gives low recoil loads and a short inboard length for vehicle mounting.

Short recoil: In short recoil weapons, the barrel only moves back far enough to allow unlocking of the breech. This is typified by the Colt .45 1911A1 pistol.

Gas operation

Gas operation is divided into three types:
Short stroke piston
Long stroke piston
Direct gas.

Short stroke piston: In the short stroke piston the gas is tapped from the barrel where it operates a small lightweight piston over a short (typically .5in) distance. This movement is transmitted to the bolt by a lightweight rod that also only moves over a short distance. This is typified by weapons such as the AR18.

Long stroke piston: Long stroke piston utilises what is, in effect, a one-piece bolt, rod and piston. This leads to a relatively heavy construction. This is typified by such weapons as the Bren gun and the AK47.

Direct gas: The direct gas system as its name implies takes the gas tapped from the barrel along the tube to the breech area where it acts directly on to the bolt. This is typified by the M16.

Locking systems

Locking systems break down into the following basic categories:
Toggle systems
Lug systems
Tilting breech blocks
Rotating bolt
Hinged or Sliding breech block.

Toggle: This system is typified by such recoil operation weapons as the Luger pistol. It is also used in such weapons as the Vickers medium machine gun. The breech is locked by the toggle being at a reverse angle and resting on the top of the barrel extension (diag 1). As the barrel extension with the toggle moves back, cams drive the toggle joint up allowing the breech block to begin to move rearward (diag 2). The toggle continues to break at its pivot point and the breech can open fully (diag 3).

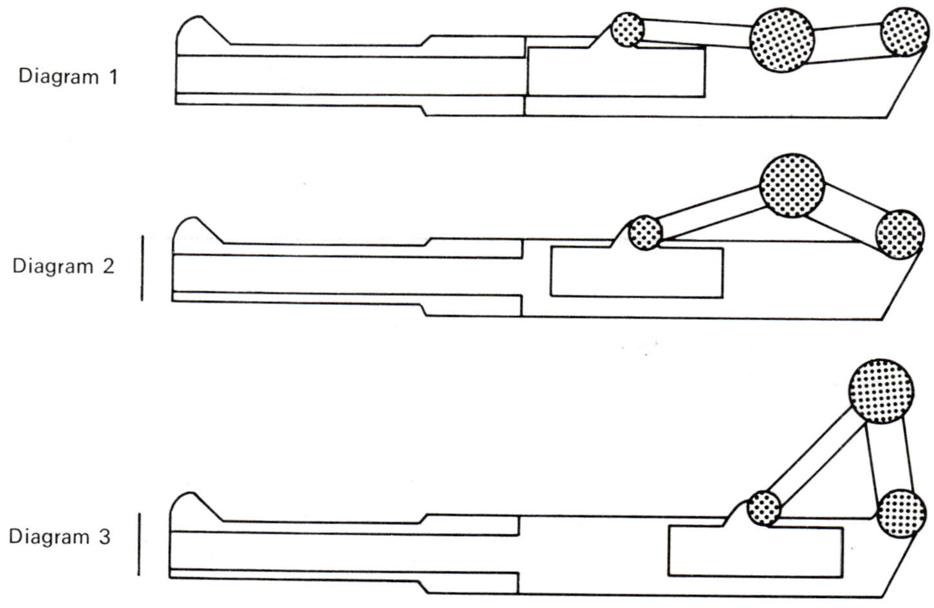

The Luger uses a similar system to the Vickers machine gun, but the toggle is cammed by ramps on the receiver

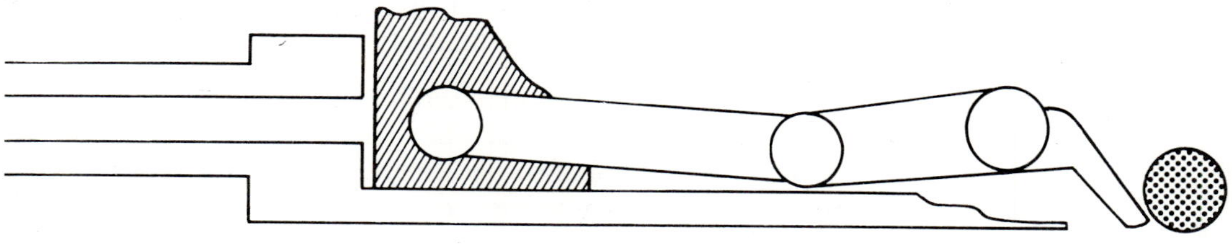

The toggle lock as used on the Vickers machine gun. The breech block is held closed by the toggle until the recoil moves the cam at the rear of the toggle against the cam pin. The cam pin drives the toggle open and this allows the breech to open

Lug system: There are a number of different types of lug locking systems. The following are examples:

- The most common form of the recoil is the link typified by the Colt 1911A1. The barrel has lugs machined into the top which lock into corresponding lugs on the slide (diag 1). When the weapon fires, the slide begins to move rearwards with the barrel locked to it. The link which is attached to the main frame of the weapon begins to pull the rear of the barrel down and thus out of engagement with the lugs. By the time the pressure is at a safe level the lugs are disengaged and the slide is free to travel to the rear (diag 2).

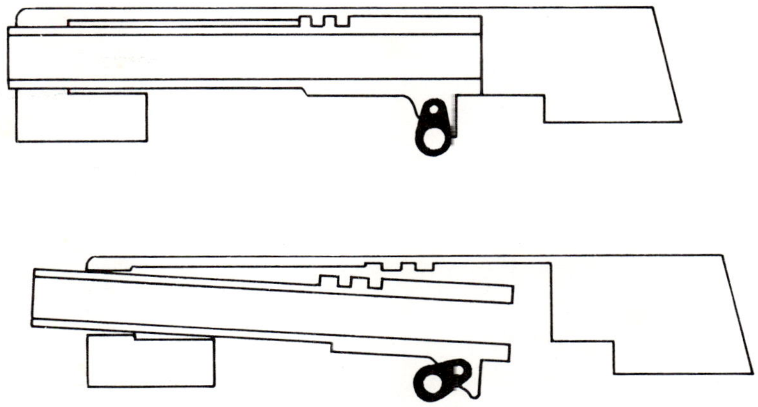

The Browning-designed swinging link system uses the recoil energy to move the slide of the pistol rearwards. This rearward motion is imparted to the barrel, because it is locked to the slide by lugs. As the barrel is also held via a link to the receiver it is pulled downwards and thus out of engagement with the lugs

- A derivation of the link type is the cam type which is used on such pistols as the Browning Hi Power and the SIG. In this, the barrel has cams on the lower side that engage with a cam pin or pins which control the movement of the barrel. The barrel is locked to the slide with the same type of lugs as on the link system.

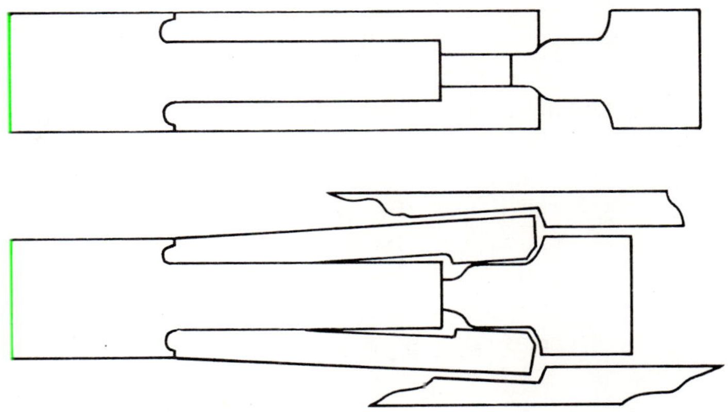

This is a form of lug lock used on the Degtyarev. It uses two struts that are cammed into locking recesses in the receiver by the striker carrier. The piston extension has cams on it that engage pins on the struts and pull them out of the receiver and allow the bolt to move rearward

- In the recoil operated P38 the movement of the slide, which moves a short distance rearward locked to the breech block, causes a separate locking piece to be cammed out of engagement. When the two lugs clear the slide it is free to recoil. This system is also used with gas control in the Czech VZ58 rifle. A further variation is typified by the Degtyarev rifle which utilises two lugs that are cammed in and out from the bolt into the sides of the main receiver, under the control of the gas system.

Tilting breech blocks: The tilting breech block system is typified by the Bren gun and utilises a breech block that pivots under the control of a cam operated by the gas system. The block moves into and out of a recess in the receiver body.

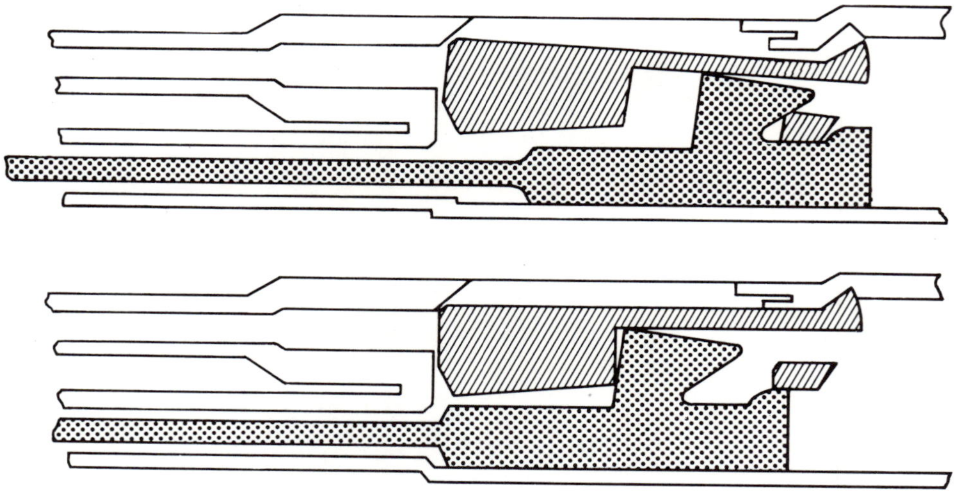

The Bren gun uses a tilting block controlled by a long stroke piston. The top diagram shows the weapon fired and the piston post moving to the rear, hitting the cam pin on the bolt and pulling the bolt down out of engagement with the locking recess in the gun body. The forward movement of the piston post drives the bolt back into the body recess

Rotating bolt: Rotating bolt typified by the AR18 and AK47 rifles is in effect a turn bolt rifle controlled by a gas system. Multi-lug bolts are rotated by cam slots in the bolt body into the barrel extension.

Hinged or Sliding breech block: This type of breech block finds little or no use on small calibre weapons and is confined to such weapons as the US M73 light machine gun.

Trends

The Russian development of small arms has been concerned initially with the production of a replacement for the AK47. This has been carried out in the adoption of the AKM. This type of change has been seen in many countries of the bloc.

At present, the Russian developments seem to be centred on two areas. The adoption of a small calibre conventional cartridge and the development of the advanced caseless type. Reports on the latter are scarce and reports of infantry trials exaggerated. Hard information on the former is available and would indicate a modified rifle, the AKD firing a new cartridge — the 5.6 mm × 45 mm. The new cartridge is based on the 7.62 mm × 39 mm case and thus the older weapons could possibly be reworked to use it. The case volume would indicate an improved performance over the basic 7.62 × 39 mm and the M193 5.56 mm all things being comparative.

Pistols and Revolvers

Most inventions, when they first appear, are looked upon with suspicion and are not often considered to be competitors to existing products. The automatic pistols such as Borchardt, Bergmann and Mauser were considered to be over-complex and were purchased by some officers more for show than for practical reasons. The Mauser 1896 'Broomhandle' was, however, an effective weapon: Winston Churchill praised it highly after he had put it to practical and successful use at the Battle of Omdurman in 1898. At the beginning of this century an American, John Moses Browning, started to design and has manufactured a series of automatic pistols and, with some models, cartridges to fit. At first these pistols were not seriously considered by the military authorities but, when it dawned on them that the automatic pistol had possibilities, trials were begun.

At a trial in the United States, in 1907, Browning submitted a number of pistol variations which culminated later in the weapon which became the basic design from which the majority of pistols up to the present date were developed.

A short recoil, locked breech weapon—this pistol was finally adopted as the Model 1911 in ·45 ACP calibre. Also entered in the trials was a ·45 calibre model of the Luger pistol. The Luger, based on an earlier American design, the Borchardt, utilised the same toggle lock and was designed to fire a 7·65 mm cartridge, from which the more powerful 7·63 mm Mauser, albeit with the same dimensions, was developed. Although unsuccessful in the United States trials, the Luger was adopted in 1908 by the German government. It was chambered for 9 mm parabellum, a cartridge that was destined to become the standard SMG and pistol cartridge for most of the world. The Luger, an expensive weapon to produce, was unreliable in adverse conditions due to the close machining tolerances which did not allow a wide enough margin to prevent it being affected by small quantities of dirt. Nevertheless, the pistol has remained in service well after production ceased in 1943.

By 1910, several countries had designed, produced and adopted a wide variety of patently unusual weapons in a selection of calibres, some firing totally ineffectual cartridges. Most of these automatics, however, were only introduced as additions rather than as replacements for the revolver.

Provided that any particular automatic pistol has proved itself to be reliable, its advantages over the revolver are fourfold. First, the automatic pistol holds more rounds than the revolver, and can be reloaded more quickly, provided that a supply of magazines is readily available. The ammunition is usually the same as that fired from a SMG. The automatic retains all the gas pressure of the cartridge, whereas in the revolver a certain amount is lost through the gap between cylinder and barrel. Most automatics fire single action having been cocked by the slide. Thus a higher rate of fire combined with greater accuracy are available which are important factors in close range combat. Lastly, pointing characteristics of the automatic are generally thought to be better than the revolver.

On the other hand, the revolver is basically accurate and reliable. It does not have to rely on the self-loading mechanism of the automatic, which requires good ammunition and a sound magazine to function correctly. It is, therefore, less prone to jamming and it is for this reason that the majority of police forces still prefer the revolver. The difficulty of accurate fire with the hand gun has led to the almost universal replacement with the

SMG, the hand gun being carried more for psychological reasons, or for personal protection, by those whose environment or work demand it. Whether relegated from front line service with the world's armies or not, the hand gun will still remain in use by many specialist units and police forces.

From the great variety of calibres used in pre-war weapons, the 9 mm parabellum has been accepted almost universally for pistol use (as well as for the SMG). However, the heavier ·45ACP is still in wide use, and is certainly a better cartridge in terms of stopping power, but requires a proportionately heavier action to handle it. The Russians, as ever, feeling the need for a round which cannot be accepted by any other than native weapons, have produced the 9 mm Makarov which is a compromise between the 9 mm parabellum and the 9 mm short.

Although modern developments have made the use of alloy frames for hand guns more prevalent, the acceptable recoil energy dictates a minimum weight.

For the future, it would appear that, unless some entirely new form of ignition is introduced, automatic pistol design will not progress a great deal, as the short recoil system has proved to be the most effective for use with a 9 mm parabellum or similar round. Thus we should look at replacing the standard cartridge with something less conventional, such as the caseless round.

Mauser pistol 1896

Luger (PO8 model) pistol

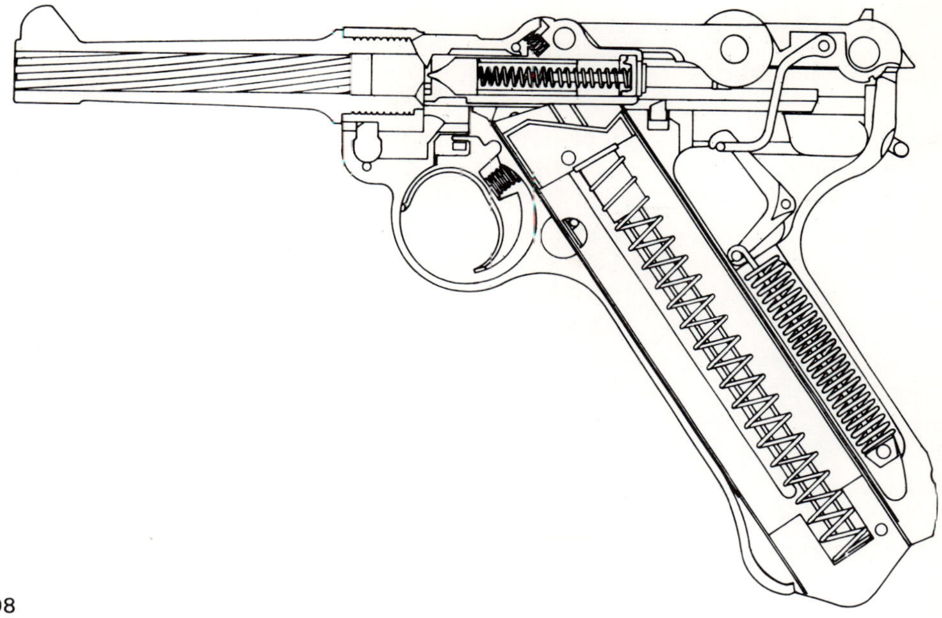

Cutaway of PO8

Vz61 Skorpion machine pistol

Czechoslovakia

Calibre	7·65 mm × 17 mm ·32 ACP
Dimensions	Length 27·6 cm stock folded 52·2 cm stock extended Barrel length 11·4 cm
Weight	1·31 kg
Effective range	50–75 m
Operation	Blowback, selective fire
Muzzle velocity	305 m/sec
Sight	Front—protected post Rear—flip over notch graduated for 75 or 150 m
Magazine	10 or 20 round detachable box type
Manufacturer	State factories

The Vz61 Skorpion is a machine pistol in the true sense. A basic blowback weapon firing the relatively weak 7·65 mm round, the Skorpion is fitted with a cyclic rate reducer. This is in fact an inertia mechanism fitted in the pistol grip. On completing its rearward stroke, the bolt strikes an actuator causing it to pivot on its pin. The actuator then strikes a spring-loaded plunger compressing the unit downwards. As the actuator pivots downwards the top part catches in the bolt thus holding it back until the spring plunger returns upwards to trip the actuator and thus release the bolt. This reduces the rate of fire from approx 1000 rpm to 800 rpm. The weapon fires from a closed bolt and has a three position selection lever giving safe, repetition and automatic. A folding wire stock should be used when the weapon is fired on automatic. The weapon was basically designed for use by armoured vehicle crews. Also available for the Skorpion is a night sighting device, a silencer, bayonet hip holster and even a shoulder holster.

In service with the Czechoslovak forces.

Vz61 Skorpion machine pistol (with stock folded)

Vz61 Skorpion machine pistol, stock extended

M52 pistol

Czechoslovakia

Calibre	7·62 mm M48, 7·62 mm × 25 mm
Dimensions	Length 21 cm Barrel length 12 cm
Weight	1 kg
Effective range	50 m
Operation	Recoil
Muzzle velocity	420 m/sec
Sight	Front—blade Rear—square notch
Magazine	8 round detachable box type
Manufacturer	State factories

Unusual in that it uses a roller locking system based on that of the German MG42, the M52 fires an unnecessarily powerful cartridge. The Czech 7·62 mm M48 round is some 20% more powerful than the Soviet 7·62 mm Type P which may also be fired in this pistol. The roller locking system, though giving great strength, is expensive to produce and rather impractical as the weapon is harder to strip than most. Unless the trigger is pulled, the weapon is prevented from firing by an automatic safety mechanism on the firing pin. The finish of the weapon is rather rough and, due to the extra power involved when using the M48 round, the gun has a heavy recoil and sharp muzzle blast. The M52 is in limited service in the Czech army as it is being replaced largely by the Vz61 Skorpion and the Wz63 machine pistols. It may, however, be encountered in Africa, Latin America and some of the Arab countries.

In service with the Czechoslovak forces and the forces of a number of African, Arab and Latin American countries.

M52 pistol

CZ75 pistol

Czechoslovakia

Calibre	9 mm parabellum
Dimensions	Length 20·6 cm Barrel length 12 cm
Weight	1·02 kg (unloaded)
Operation	Blowback—double action
Muzzle velocity	355 m/sec
Sight	Front—blade Rear—notch
Magazine	15 double column
Manufacturer	State factories

Unlike most of the Warsaw Pact hand guns, the CZ75 is of extremely high quality, design and finish. It combines the best of the Browning HP and the Swiss SIG210 pistols. It fires both double and single action and has a magazine capacity of 15 rounds. Being made in 9 mm and 19 mm calibre suggests that this pistol was intended for the export market.

CZ75 pistol

Pistole M pistol

Germany (DGR)

Manufacturer State factories

See the Makarov.

In service with the DGR forces.

Walther PP and PPK pistols

Germany (FGR)

Calibre	·22LR, 6·35 mm/·25ACP, 7·65 mm/·32ACP, 9 mm Short
Dimensions	Length 17·2 cm Barrel length 10 cm
Weight	0·7 kg
Effective range	35 m
Operation	Blowback
Muzzle velocity	290 m/sec (7·65 mm)
Sight	Front—blade Rear—U notch
Magazine	8 round detachable box type
Manufacturer	Carl Walther Waffenfabrik

The Walther PP and its smaller edition the PPK, were first introduced in 1929 and 1931 respectively and were among the first automatic pistols to feature a double action trigger mechanism. The pistols were well designed and manufactured with simple blowback actions. Featuring not only a double action trigger, but also a loaded chamber indicator which protruded out of the rear of the slide, the PP and PPK have not only been sold in their more common calibres of 7·65 mm and 9 mm short but also in ·22 in and 6·35 mm. Used in large quantities by the German Army and Police during World War II, the Walther pistols are being mass produced again and have sold widely abroad, especially to police forces. France, Hungary and Turkey have copied or produced the weapons under licence and a number of countries have manufactured copies based almost exactly on the originals.

In service with the French, German, Hungarian and Turkish forces.

Walther PP pistol

Walther PPK pistol

48M pistol

Hungary

Manufacturer State factories

See the Tokarev TT33.

In service with the Hungarian forces, but obsolescent.

PA-63 pistol

Hungary

Calibre	9 mm, 7·65 mm
Dimensions	Length 17·5 cm
	Barrel length 8·7 cm
Weight	0·66 kg (loaded)
Effective range	50 m
Operation	Blowback
Muzzle velocity	315 m/sec
Sight	Front—blade
	Rear—notch
Magazine	7 round detachable box type
Manufacturer	State factories

The 9 mm pistol PA-63 is the latest of several lightweight pistols to be introduced into the Hungarian armed forces, all resembling to some extent the German Walther PP pistol. A standard blowback-operated, double action weapon with very little bulk, the PA-63 is light and manageable for close quarters work, but the claimed effective range of 50 m appears to be optimistic and the small capacity of the magazine is a disadvantage. There are several slightly different versions of the PA-63, varying mainly in the material of the grip and the body finish, but all carry the same thumb-operated safety catch on the left side near the external hammer and on the same side the magazine release catch forward of the top of the grip.

In service with the Hungarian forces.

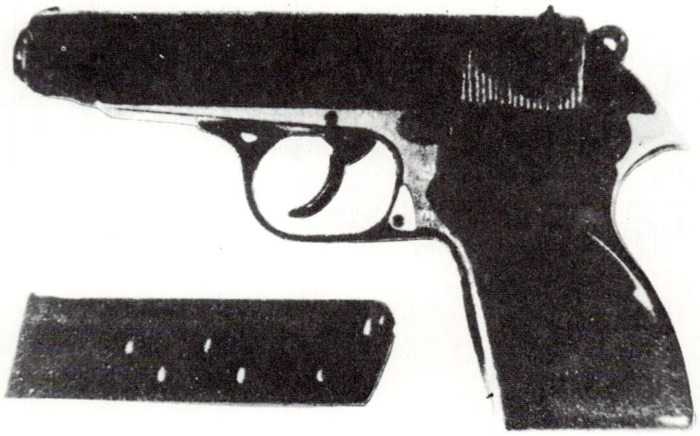

PA-63 pistol

Pistolet TT pistol

Poland

Manufacturer State factories

See the Tokarev TT33.

In service with the Polish forces, but obsolescent.

P64 pistol

Poland

Calibre	9 mm Makarov
Dimensions	Length 15·5 cm
	Barrel length 7·9 cm
Weight	0·596 kg
Effective range	35 m
Operation	Blowback
Muzzle velocity	305 m/sec
Sight	Front—blade
	Rear—notch
Magazine	6 round detachable box type
Manufacturer	State factories

The P64 lightweight pistol now replaces the old Pistolet TT in the Polish army. The pistol outwardly resembles the Makarov and has some features copied from the Walther PP. A double action weapon, the trigger mechanism differs from either the Makarov or Walther designs. An internal slide stop is fitted and the safety catch is situated at the rear of the slide on the left-hand side.

In service with the Polish forces.

Radom VIS 35 pistol

Poland

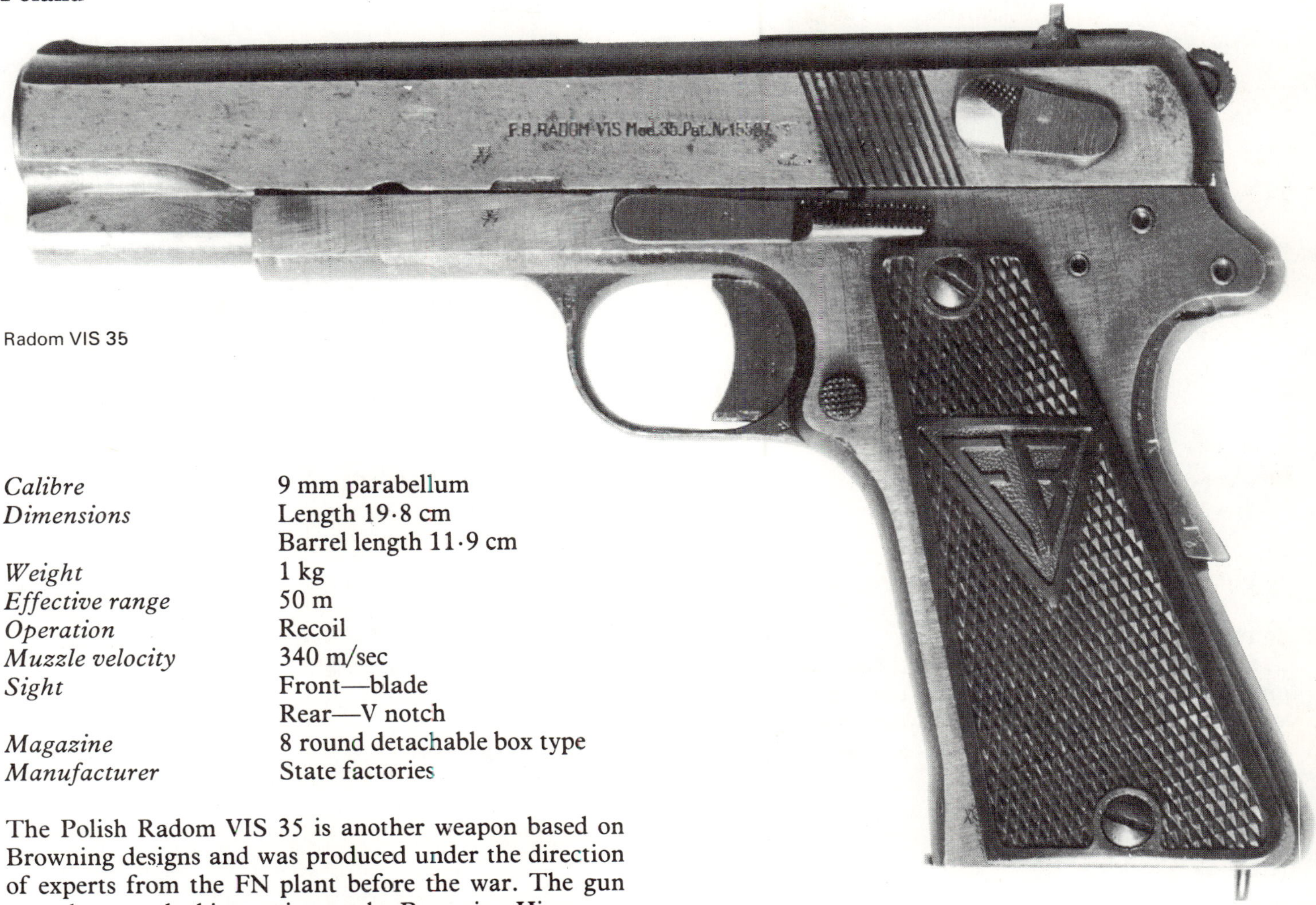

Radom VIS 35

Calibre	9 mm parabellum
Dimensions	Length 19·8 cm
	Barrel length 11·9 cm
Weight	1 kg
Effective range	50 m
Operation	Recoil
Muzzle velocity	340 m/sec
Sight	Front—blade
	Rear—V notch
Magazine	8 round detachable box type
Manufacturer	State factories

The Polish Radom VIS 35 is another weapon based on Browning designs and was produced under the direction of experts from the FN plant before the war. The gun uses the same locking action as the Browning Hi-power. As the weapon was designed with the cavalry in mind an unusual device was fitted. The lever on the slide on being pushed downwards first blocks the firing pin and then lowers the hammer. Two slide catches are fitted, the forward one being a hold-open device for when the last shot is fired. The rear one (mounted in the same place as the safety catch on the 1911A1) is merely a slide hold to assist in dismantling. A grip safety is also fitted. The VIS 35 is a robust weapon and can handle all types of 9 mm round.

In service with the Polish forces.

Wz63 machine pistol

Poland

Calibre	9 mm Makarov, 9 mm × 18 mm
Dimensions	Length 33·3 cm (with butt extended 60·7 cm)
	Barrel length 15 cm
Weight	1·8 kg (loaded)
Effective range	50 m (200 m with butt extended)
Operation	Selective fire, blowback operated
Muzzle velocity	325 m/sec
Sight	Front—blade
	Rear—protected notch
Magazine	25 or 15 round detachable box type
Manufacturer	State factories

The Polish Wz63 machine pistol was produced to utilise the new Soviet 9 × 18 mm round and follows the same design ideas as the Soviet Stechkin (APS) and Czech Skorpion machine pistols. While the butt and forward grip are still folded, the weapon can, in spite of its ungainly appearance, be used accurately as a normal pistol and is usually carried in a holster. To be used as a sub machine gun, the forward grip should be lowered until it locks at 90° to the barrel and the butt, normally folded

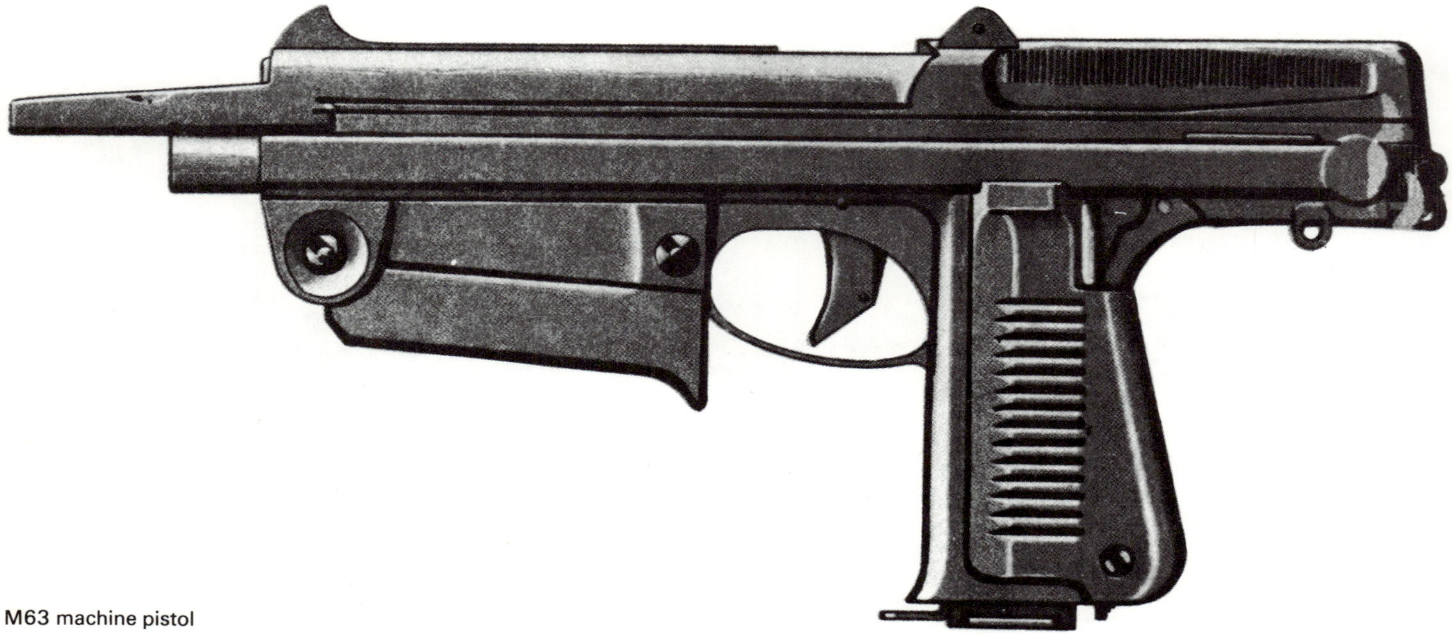

M63 machine pistol

on either side of the pistol body, should be retracted to the rear and the butt plate revolved into position. If required, the attached leather strap can be used as a brace support. The Wz63 fires from the open bolt position and fire is selected by the different finger pressure on the trigger, not the selector switch. The distinctive semi-circular extension below the muzzle is a compensator to reduce the amount of recoil when firing on automatic. The smaller 15 round magazine is designed to be used when the weapon is employed purely as a pistol. The main purpose of the Wz63 is to act as the personal weapon of troops operating in a very confined space such as an armoured vehicle. It is standard issue to Polish paratroops.

In service with the Polish forces.

Stechkin (APS) pistol

USSR

Calibre	9 mm Makarov, 9 mm × 18
Dimensions	Length 22·5 cm
	Barrel length 12·7 cm
Weight	1·03 kg
Effective range	50 m without stock
	150 m with stock
Operation	Blowback
Type of fire	Selective
Muzzle velocity	340 m/sec
Sight	Front—blade
	Rear—notch
Magazine	20 round detachable box
Manufacturer	State factories

Stetchkin (APS) pistol

This pistol is a true machine pistol in that it is selective fire. This feature, even with the low powered cartridge, is not efficient as the weapon is very difficult to control. The fitting of the shoulder stock makes accurate fire possible at short ranges. The Stechkin is no longer in front line service as it has been superseded by the Makarov.

In service with USSR border guards.

Tokarev TT33 pistol

USSR

Calibre	7·62 mm type P
Dimensions	Length 19·3 cm Barrel length 11·4 cm
Weight	0·8 kg
Effective range	50 m
Operation	Recoil
Muzzle velocity	420 m/sec
Sight	Front—blade Rear—U notch
Magazine	8 round detachable box type
Manufacturer	State factories

The Tokarev pistol first appeared during the early 1930s. Again of the basic Colt-Browning design, the TT33 has some interesting features. A removable sub-assembly block containing the hammer mechanism and forming part of the frame has its inside surfaces machined to act as a cartridge guide. Most feed malfunctions are due to weak or damaged magazine lips but this feature of the TT33 almost eliminated this problem. The gun has no safety catch but the slide and trigger may be locked by a half-cock position on the hammer. The barrel locking lugs run the complete circumference of the barrel for ease of manufacture on later models.

Used extensively by Soviet forces during World War II, the Tokarev has been produced throughout the Warsaw Pact countries and China. Hungary manufactured a version for Egypt called the Tokagypt in 9 mm parabellum. The Tokarev types differ only in the motif on the grips and the slide serrations.

In service with the Warsaw Pact and other communist forces, but obsolescent.

Tokarev TT33 pistol

Makarov PM pistol

USSR

Calibre	9 mm Makarov
Dimensions	Length 16 cm Barrel length 9·6 cm
Weight	0·8 kg (loaded)
Effective range	35 m
Operation	Blowback
Muzzle velocity	420 m/sec
Sight	Front—blade Rear—square notch
Magazine	8 rounds detachable box type
Manufacturer	State factories

Although based on the Walther PP, the Makarov (PM) pistol differs in some respects. The main differences are an externally mounted slide stop, a spring type magazine catch and the lack of a loaded chamber indicator. The trigger is a double action type.

Now standard in most major Euro-Asian communist armies, the Makarov is also manufactured in East Germany as the Pistole M (illustrated) and in China as the Type 59 pistol. The Soviet Makarov carries a star in the centre of each side of the grip.

In service with the Warsaw Pact and other communist forces.

Makarov PM pistol

M57 and M65 pistols

Yugoslavia

Manufacturer	State factories

See the Tokarev TT33.

The M65 is a 9 mm parabellum version of the M57.

In service with the Yugoslav forces, but obsolescent.

M67 pistol

Yugoslavia

Calibre	7·65 mm/·32 ACP, 9 mm short/9 mm × 17 mm
Dimensions	Length 16·5 cm Barrel length 9·4 cm
Weight	0·8 kg (loaded)
Effective range	50 m
Operation	Blowback operated, semi-automatic only
Muzzle velocity	310 m/sec
Sight	Front—blade Rear—notch
Magazine	6 round detachable box type
Manufacturer	State factories

The Yugoslav M67 pistol is the latest to be added to the wide variety of pistols used by the armed forces of that country. It is manufactured in two different calibres to take the 9 × 17 mm and the 7·65 × 17 mm SR rounds. The M67 is a very small, lightweight pistol with only a limited magazine capacity, weighing only 0·7 kg empty. The safety catch is located, for thumb operation, in the cut-away angle of the top of the left-hand grip, while the magazine release catch is also on the left of the grip behind the base of the trigger guard. Unlike most pistols produced by the Warsaw Pact countries, the slide assembly is retained and released by a similar slide stop unit and pin to that of the 9 mm Browning pistol located on the left above the trigger.

In service with the Yugoslav forces.

Sub machine Guns

Although the earliest SMG design dates back to the 1890s, the first practical examples did not appear in any numbers until after the start of World War I. In April 1914, an Italian named Bethel Revelli took out a patent on a weapon originally designed for use in aircraft. First produced at the Villar–Perosa plant, and later by Fiat and under licence for the Italian Government in Canada, the Villar Perosa (VP) as it is universally known, was later adapted for use by infantry and was mounted on motor cycles and bicycles. The VP consisted basically of two weapons held together by a crossbar and used a form of retarded blowback in that the bolt was cammed to turn slightly during the forward motion. Either mechanism could be fired individually, or both together, using 25 or 50 round magazines. A mounting was also available by which the weapon could be slung around the firer's neck and operated on the move.

At the battle of Caporetto in 1917, large numbers of VPs were captured by the Germans who sent a number back to Germany for evaluation. Earlier in the same year, the Germans, realising the need for a light automatic weapon to fill the gap between rifle and pistol, had begun issuing larger numbers of the 7·63 mm and 9 mm Mauser pistols, the holsters of these weapons being able to double as a shoulder stock. In addition, production of a longer barrelled version of the P08 Luger pistol was increased. Fitted with a tangent rear sight and optional stock, the artillery model, as it was known, could also be fitted with a specially designed and unique 32 round snail drum magazine which gave it considerably increased firepower. Although this magazine was awkward and required a loading tool to fill it to capacity, the weapon did go some way towards filling the basic requirements for a sub machine gun. It was lighter and more compact than a rifle, and although firing the same cartridge as the pistol, gave considerably more firepower.

The realisation of the usefulness of this type of weapon led to the acceleration of design work started in 1916 by Hugo Schmeisser, later joined by Theodore Bergmann. By June 1918, the Bergmann MP18:1 was being issued to front line units. The MP18:1 was a basic blowback weapon which also took the 32 round snail drum magazine, though this had to be fitted with a removable sleeve to allow for the shorter magazine housing of the SMG. The gun fired full automatic only, had a wooden stock, a ventilated barrel jacket and at 4·1 kg was heavier than the standard rifle.

The Treaty of Versailles required, inter alia, that the German Army should not hold any SMGs and the MP18:1 was relegated to use by the police. The gun was later modified by changing the magazine housing to take a straight box type magazine, and provision was made for firing single shots. This was the MP28:11.

The Americans, likewise feeling the need for a weapon in the hands of an infantryman which was capable of putting down concentrated fire, began to inaugurate ideas for a truly unique system, though this cannot really be classed as a SMG. Rather than go to the lengthy and expensive process of designing a special weapon, a unit was produced which could be fitted to the M1903 Springfield rifle. Known as the Pedersen device (official secrecy shrouding it under the designation 'Automatic Pistol Cal ·30 Model 18'), the unit could be inserted into the rifle in place of the turn bolt. A 40 round magazine containing ·30 pistol type cartridges was then inserted and the blowback operated action fired the weapon semi-automatically.

Although 500,000 Model 18s were eventually ordered, only 65,000 units were produced and none saw service as it was felt that the extra weight (1 kg and 10 magazines at 0·5 kg each), together with the likelihood of the soldier losing the ordinary bolt, considerably outweighed any advantages.

John Taliaferro Thompson's famous series of SMGs began in 1917 with the belt fed persuader in ·45ACP. Using a retarded blowback system which incorporated the Blish hesitation lock (a sliding H-piece straddling the bolt to retard it until the pressure dropped), the gun failed miserably. However, as developments continued, so matters improved, and the Annihilator fitted with a box magazine reached prototype form just too late to be of use to the US forces in World War I. The first production model appeared in 1921 and, though the tests by the US Marine Corps were favourable, the gun did not sell. The 1921 model featured a choice of box or drum magazines, the latter being dispensed with when the weapon was modified for pure military use to become the M1 in 1942. By the time the M1 was adopted, it had been proved that the Blish piece was unnecessary and, along with several other expensive features, was done away with thus making the gun pure blowback-operated.

In the period between the wars, several countries experimented with SMG designs, but few realised the potential of this type of weapon. Consequently, experimentation and trial costs were mostly kept to a minimum. Blowback operation was now being accepted as the most efficient system although some other types had been introduced including a toggle action by Furrer.

Having seen the effects of SMG use at first hand during the Spanish Civil War, the Germans then went on to design and manufacture their MP38 prior to World War II. This weapon was the first SMG produced with the salient features now incorporated in most modern SMGs. A simple turned bolt, combined with a steel tube receiver with exterior milled slots for lightness, made the gun relatively easy to produce. The firing pin and operating spring were manufactured as a telescoping unit preventing spring kinkage and the entry of dirt. The most revolutionary feature, however, was the fitting of a folding stock, the gun being primarily designed for use by parachute and armoured troops which required the gun to be stored and operated in confined spaces. Although no fire selection button was fitted, the weapon had a low enough rate of fire to permit single shots to be tapped off by a trained operator. The later version, MP40, cut down production costs even further by using a stamped steel receiver and by the modification of other parts. The magazines were also strengthened by the addition of two ribs to either side.

Both the British and Americans, slow to realise the potential of a new arm, began ordering vast numbers of the heavy and expensive Thompsons at the start of World War II, even though the Americans had adopted the weapon in 1938.

Finally, the British began to design their own SMG. Deciding to make their own copy of the German MP28:11, manufacture of the Lanchester began in early 1941, later being issued to the Royal Navy. At the same time, it was felt that a lighter and cheaper model was required and this led to the Sten gun. Manufactured using a maximum of stampings and spot welding, the Sten in its various marks went on to become the cheapest SMG ever put into mass production. The stock was removable and thousands, including silenced versions, were issued to Allied soldiers or dropped to resistance fighters. The Germans even produced a copy of the Sten during the closing months of the war.

The Russians, showing slight pre-war interest in the SMG, had also manufactured a copy of the German MP28:11 but in 7·62 mm Type P as opposed to the 9 mm parabellum of the original. The change in calibre was due to the fact that the Russian barrel makers had been producing the 7·62 mm Nagant rifle for many years and the adoption of the same bore size did not necessitate re-tooling. Their later PPS and PPSH designs were cheap and easy to produce and featured a 71 round drum magazine giving greater firepower, but they were difficult to load at night and not very rugged. The Russians became the largest users of the sub machine gun, even going to such lengths as equipping special squads to ride on the back of tanks armed only with SMGs.

The advantages of the Sten and similar weapons, in terms of production costs, soon became apparent to the Americans who proceeded to design one capable of being manufactured with similar ease and having the advantage of firing either the ·45ACP or the 9 mm parabellum cartridges with a minimum change of parts. Designated the M3, the weapon was popularly known as the Grease Gun and first appeared in December 1942.

Today's weapons are basically only refinements of the designs laid down 30 or 40 years ago, though in peacetime the standard of manufacture tends to rise.

The most advanced feature to come out of post-war SMG design is the wrap-around bolt. This type of bolt has the advantage, not only of allowing the weapon to have a much shorter and compact receiver, but also of displacing the weight of the bolt more towards the muzzle. This eliminates a great deal of muzzle climb on full automatic fire and thus reduces the need for a compensator, previously the accepted way of reduction. In general these weapons are also more accurate.

With few exceptions, the universally accepted SMG cartridge is the 9 mm parabellum. The drum magazine has now been completely superseded by the box type which, on some weapons, is located inside the pistol-grip most notably in the wrap around bolt type guns. The box magazine has advanced considerably and with most modern types a loading tool is no longer necessary. Nonetheless the Warsaw Pact countries, and particularly Russia, have phased out a great number of their SMGs in preference to their assault rifles which they use as both rifle and SMG. The future of the SMG within NATO, however, like the LMG, is now dependent on the new calibre which is yet to be decided. If, as in the case of the USA, the calibre is to be of 5·56 mm or similar, then it is likely that the SMG will be gradually phased out in preference to a dual purpose weapon, or system, except for special operations. Whatever the case, the SMG will be found in use for many years to come in many armies, and particularly by revolutionary and similar groups or by police forces.

It is worthy of note that, since the SMG first appeared, almost every country with the engineering capacity has produced a SMG of some description, though the majority never progressed beyond the prototype stage.

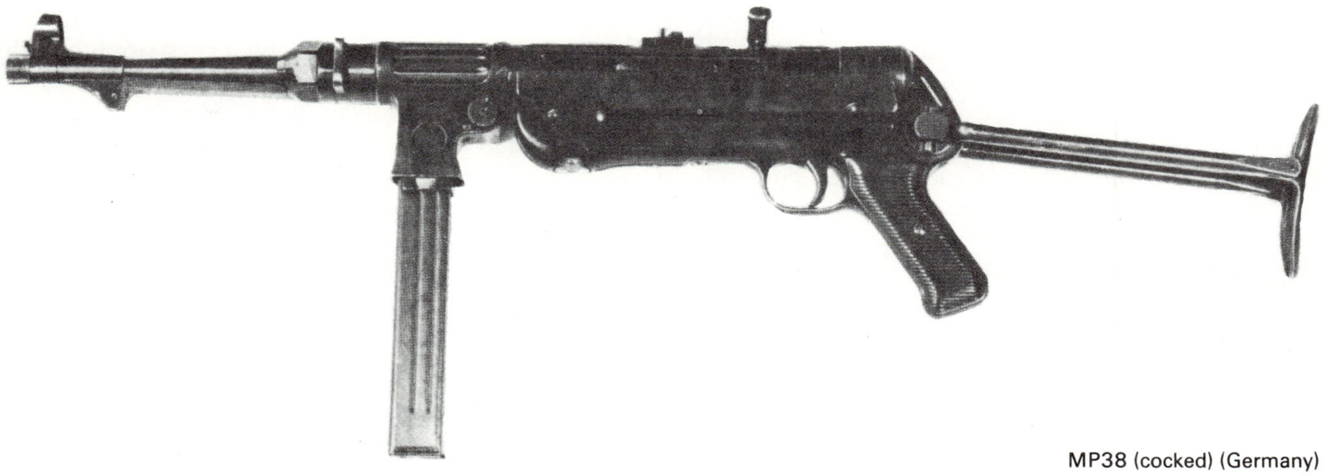

MP38 (cocked) (Germany)

Vz23 (Czechoslovakia)

Vz23 and Vz25 sub machine guns

Czechoslovakia

Calibre	9 mm parabellum
Dimensions	Length 68·6 cm (stock extended), 44 cm (stock folded)
Weight	3·1 kg (Vz23 without magazine)
Effective range	200 m
Operation	Blowback
Type of fire	Selective
Muzzle velocity	380 m/sec
Sight	Front—hooded barleycorn Rear—rotary V notch
Magazine	24 or 40 round detachable box type
Cyclic rate	600 rpm
Manufacturer	State factories

The Czech Vz23 and Vz25 are blowback-operated weapons, the distinction being that the Vz23 has a fixed stock and the Vz25 a folding stock. Like the Uzi, the barrels of the Vz23 and Vz25 are extended well into the receiver and the recessed bolt 'wraps around' the barrel in the forward position. An ejection port is cut out of the side of the bolt, thus the receiver ejection port is closed to dirt whether the weapon is cocked or not. The butt-piece of the Vz25 may be used as a fore pistol-grip when the stock is folded. The pistol-grip, like the Uzi, also serves as a magazine housing. A magazine-loader is attached to the right-hand side of the foregrip and is used with 8 round stripper clips. To prevent the weapon from accidentally discharging should it be dropped, the cocking-handle is locked into its slot in the receiver by a tilting process, thus the handle may only be withdrawn manually. Now obsolete in the Czech army, large numbers were sold to Syria and Cuba.

In service with the Cuban and Syrian forces.

Vz23

Vz25, stock folded

Vz25, stock extended

M24 and M26 sub machine guns

Czechoslovakia

Calibre	7·62 mm M48
Dimensions	Length 67·5 cm (wooden stock), 68·6 cm (metal stock extended), 44 cm (stock folded) Barrel length 28·4 cm
Weight	4·2 kg (M26 with loaded magazine)
Effective range	200 m
Operation	Blowback
Type of fire	Selective
Muzzle velocity	550 m/sec
Sight	Front—hooded barleycorn Rear—rotary V notch
Magazine	32 round detachable box type
Cyclic rate	600 rpm
Manufacturer	State factories

The current SMGs of the Czech army are the M24 and M26 which are very similar to the Vz23 and Vz25, except for a change of calibre to 7·62 mm M48, the powerful Czech pistol cartridge. The main exterior difference is that the pistol-grips of the M24 and M26 point slightly forward and a phosphated finish is applied as opposed to the black finish of the Vz23 and Vz25.

In service with the Czechoslovak forces.

M48 sub machine gun

Hungary

Manufacturer	State factories

See the PPSh41.

Obsolescent.

PPSh41 sub machine gun

USSR

Calibre	7·62 mm type P, 7·62 mm × 25 mm
Dimensions	Length 82·2 cm Barrel length 26·5 cm
Weight	3·9 kg
Effective range	150 m
Operation	Blowback
Type of fire	Selective
Muzzle velocity	500 m/sec
Sight	Front—hooded post Rear—tangent leaf or L type with U notch
Magazine	71 round drum or 35 round box type
Cyclic rate	700–900 rpm
Manufacturer	State factories

Designed by George Shpagin in 1940, the PPSh41 met the need for an easily mass produced, but rugged, SMG.

The receiver and barrel-jacket are stamped out, the main milled parts being the bolt and the barrel, the bore of which is chrome-plated. The PPSh41 has a select fire lever positioned in front of the trigger and the rather elaborate tangent sight on the early models was replaced by a simplified L type. The two-piece bolt handle allows the bolt to be locked in either the forward or the rear position. A 71 round drum, or a 35 round box, magazine may be fitted, the former giving the gun a great advantage over its German World War II counterpart, the MP40. Moreover, the Germans even went so far as to convert a number of captured PPSh41s to 9 mm parabellum. Although the PPSh41 is obsolete in the USSR and Soviet bloc, a large number is probably being held in reserve as there were over 5 million produced during World War II alone.

The weapon has also been produced in large numbers in China as the Type 50 SMG (taking only the box magazine), North Korea as the Type 49 (drum magazine only), Hungary as the M48 and in Iran as the M22.

A North Vietnamese version known as the K50M is merely a Chinese Type 50 with the barrel-jacket shortened and the wooden stock replaced by a pistol-grip and metal housing incorporating a wire stock similar to that used on the French MAT49.

Obsolete, but in service in various parts of the world (see text).

PPSh41, with 71 round drum magazine

PPS43 (Sudayev) and 43/52 sub machine guns

USSR

Calibre	7·62 mm Type P, 7·62 mm × 25 mm
Dimensions	Length 82 cm (stock extended), 61·5 cm (stock folded) Barrel length 23 cm
Weight	3·6 kg
Effective range	150 m
Operation	Blowback
Type of fire	Automatic
Muzzle velocity	490 m/sec
Sight	Front—post with ears Rear—L type
Magazine	35 round detachable box type
Cyclic rate	650 rpm
Manufacturer	State factories

Primarily made of sheet metal stampings, the PPS43 is a

modified version of the PPS42 which was designed, built and field-tested during the siege of Leningrad.

Cheaper to produce than the PPSh41, the PPS43 fires fully automatic only and has no fixed ejector. The main operating spring guide-rod acts as an ejector when the bolt is in the rearward position. The 35 round curved box magazine, although similar to that of the PPSh41, is not interchangeable. The stock is unusual in that, when folded, it lies on top of the receiver.

Although obsolete in the Soviet army, the PPS43 is still in use in Poland as the model 43/52 and fitted with a wooden stock. Also produced in China as the Type 43 copy.

In service with the Chinese, North Korean, Polish and other Warsaw Pact sponsored forces.

PPS43, stock folded

PPS43

M56 sub machine gun

Yugoslavia

Calibre	7·6 mm Type P, 7·62 mm × 25 mm
Dimensions	Length 87 cm (stock extended), 59 cm (stock folded) Barrel length 24·7 cm
Weight	3 kg
Effective range	75 m
Operation	Blowback
Type of fire	Selective
Muzzle velocity	520 m/sec
Sight	Front—hooded blade Rear—L type U notch
Magazine	32 round detachable box type
Cyclic rate	570–620 rpm
Manufacturer	State factories

Introduced to replace the earlier M49 (basically a copy of the Soviet PPSh41), the M56 uses the best points of the M49 and the German MP40.

The lower body, pistol-grip and folding stock are almost direct copies of the MP40, but the bolt is very similar to that of the M49. The M56 is a select fire weapon and may be fitted with a knife bayonet. The magazine is thought to be interchangeable with that of the M49.

In service with the Ethiopian and Yugoslav forces.

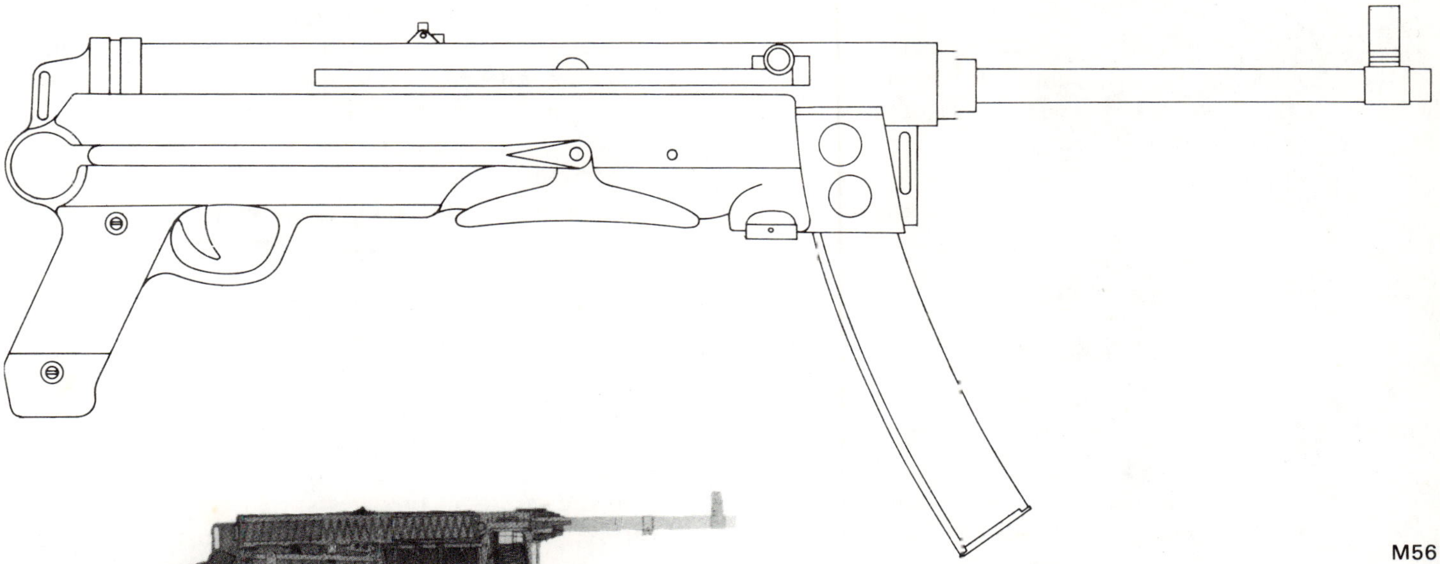

M56

Cutaway drawing of the M56

Rifles and Carbines

Serious discussion on rifles continues to be hinged around the question of the bullet and its calibre. There is, of course, the overriding need to reduce the infantryman's load, yet the rifle must be able to give accurate fire out to 400 m. A small calibre cartridge will give this performance and enable the rifle to be light in construction and the ammunition to be carried in greater quantities. This will help to give the soldier a weapon that is easy to handle in the confines of an armoured personnel carrier or operate in the urban or rural jungles—in other words, a rifle of the light carbine type. On the other hand, the argument continues, the bullet must be capable of being fired from the LMG and remain lethal at ranges up to 800 m, and there lie the conflict and dilemma.

During World War II, both Germany and Russia developed an intermediate cartridge, described as being midway between a rifle and a sub machine gun (SMG) round. Both countries used their standard rifle bullet size with a smaller case. The Russian cartridge (designated 7·62 mm × 39 mm M43) was first used in the SKS rifle, and the round remains standard today in the AK47, AKM and the earlier model LMGs, although it is worthy of note that the new PK general purpose machine gun (GPMG) returns to the old rimmed 7·62 mm M1908 cartridge, which gives ranges in excess of 900 m.

When in 1945 the British set up their ideal small arms calibre panel, they were under the impression that they would be developing information mutually with the USA. In 1947, when they visited the USA with the results of their tests, they learned for the first time of the US T65 cartridge and the T25 rifle. The competitive demonstrations in 1950 between the T65 cartridge (later destined to become the 7·62 mm (NATO) round) and the British ·280 in or 7 mm round were held using the US T25 rifle and the British designed EM2 (Bull pup) rifle. With a muzzle velocity of 700 m/sec (comparing well with the Soviet M43 round at 675 m/sec), the early loadings of the British cartridge were somewhat handicapped when pitted against the 30–06 ballistics of the T65. Yet, although the British EM2 rifle outperformed the T25, the excellent performance of the ballistically superior ·280 in cartridge at longer ranges went unnoticed. Finally, due to overwhelming pressure from the USA, the T65 was adopted as NATO standard in December 1953 and the British finally abandoned their EM2 rifle and ·280 in cartridge in favour of the 7·62 mm × 51 mm cartridge and the Belgian FN rifle. The 7·62 mm cartridge has since remained the standard round for all NATO countries; the French still retain their 7·5 mm round for their standard rifle the M49/56 and their AAT 52 LMG.

The subsequent adoption by the USA of the 5·6 mm round was in direct contrast to their earlier insistence on a full rifle calibre, in preference to an intermediate type originally proposed and in use by the Soviet block, China and other European countries, and caused a great deal of controversy. This very high velocity round was developed by the Armalite Company in 1957. It has an approximate muzzle velocity of 980 m/sec (using the standard AR18) and the makers claim that: 'Impact results in a hydrostatic effect and corresponding extensive rupture of flesh in the immediate bullet path and displacement of fluids which are driven into body tissue over a wide area. This causes a major shock and a high percentage of fatalities when compared with any present 7·62 mm (·30 cal) service load'. At ranges exceeding 550 mm, however, when the round has decelerated appreciably, the velocity effectiveness is almost lost. Many military commentators consider, however, that 350 m–400 m is the maximum killing range required of a military rifle, and so the 5·56 mm has much in its favour, especially as a rifle produced in this calibre need not normally weigh more than 3 kg. More ammunition may be carried by the soldier and, if using a breech-locking system similar to that of the M16 or AR18, the weapon becomes cheaper and easier to produce.

Initially problems were experienced with the 5·56 mm round in the M16 rifle. These were primarily caused by an unannounced change in propellant which caused an excessively high rate of fire and a theory that the M16 was self-cleaning. A number of modifications to the weapon along with introduction of normal cleaning procedures has led to the M16 becoming a reliable weapon. The 5·56 mm round has proved satisfactory up to ranges of 400 m and as a result many countries are experimenting with calibres between 4·5 and 5 mm for their next generation of weapons.

Indeed, most major European arms companies are experimenting with, or have produced rifles for sale in 5·56 mm, eg FN, Beretta, SIG, H & K and MAS. None of these weapons has been sold in large numbers but, if NATO adopts the 5·56 mm cartridge, most countries will probably equip with their own native-designed weapons. Most of these rifles are capable of launching grenades and taking a bayonet, the retention of which appears to be more of a tradition and for ceremonial than as a practical requirement, except as a knife or combination wire cutter. A folding or telescopic stock version is also offered for use by parachute troops and for carriage in helicopters and armoured vehicles. Extras include a three round burst capability, telescopic sights, night sights and bipods.

Many military experts and experienced infantrymen feel that an intermediate cartridge such as the late lamented British ·280 in would be the answer for the next generation of small arms. This would enable one cartridge to carry out all the needs of the rifle, light machine gun and GPMG. Nevertheless, the fact remains that the 7·62 mm NATO round will, because of the sheer volume of weapons and ammunition available, remain as the general purpose machine gun cartridge. NATO will, however, adopt a new calibre although whether this will be the 5·56 mm or one of the many smaller rounds in development is still open to conjecture. This cartridge will attempt to provide for light machine gun use out to 800 m as well as rifle use out to 400 m.

Since the end of World War II, arms manufacturers have made increasing use of steel stampings wherever possible. With their obvious cost and weight advantages, they enable some of the simpler types to be produced under licence in countries lacking the more sophisticated types of milling, cutting and boring machines. The wooden stock, so long a standard material in most arms, is now disappearing in favour of the plastic or composition type. In comparison with cost of replacement or the time-consuming repair of a cracked or chipped stock instead, the plastic stock needs merely to be replaced. Moreover, during the Vietnam war it was found that the wooden stocks of the M14 rifles were becoming so warped as to make stripping almost impossible.

Nevertheless, those who discard the ability to kill at longer ranges do so at their peril. At frequent intervals the value of good shots who are able to hit targets at 700 to 800 m has been forgotten. Time and again a sharp reminder has been received when actual combat occurs—even in urban areas. Certainly, snipers are invaluable at all times, but the skill of firing at longer ranges should not be lost—especially to the NATO infantry. (See also Introduction.)

In most countries, the sniper rifle has merely meant the 'accurizing' of an old standard bolt-action rifle with the addition of a telescopic sight. However, the Russians have gone against this trend by introducing the semi-automatic Dragunov. Using the basic Kalashnikov action, this rifle, specifically designed for sniping, is chambered for the old, but well proven, 7·62 mm M1908 rimmed cartridge which, as previously stated, has also been reintroduced for the PK GPMG. The Dragunov also features an unusual skeleton stock with pistol grip. The rifle was introduced to replace the M1890/30 Moisin Nagant sniper model, although this rifle is probably still in second line use by Warsaw Pact countries.

The French also have a special sniper rifle, the FRF1. Although a conventional bolt-action type, it was specially designed and features a folding bipod and a stock, the length of which may be altered to suit the individual. Some countries such as Belgium and Germany merely add a telescopic sight to their standard rifle although they undoubtedly hold stocks of bolt-action weapons in reserve.

Vz58 rifle

Czechoslovakia

Calibre	7·62 mm × 39 mm
Dimensions	Length 84·3 cm (Vz58p), 63·5 cm (Vz58v stock folded)
	Barrel length 40 cm
Weight	3·1 kg
Effective range	400 m
Operation	Gas
Muzzle velocity	700 m/sec
Sight	Front—protected post
	Rear—tangent leaf
Magazine	30 round detachable box type
Cyclic rate	700–800 rpm
Manufacturer	State factories

Replacing the earlier M53/57, the Czech Vz58 is externally similar to the Soviet AK47 though it differs internally in several aspects.

The gas system utilises a short-stroke piston to drive the action rearwards, whereas the piston and bolt carrier of the AK47 are machined in one piece. There is no facility for gas adjustment. The action is locked by a tilting locking piece, whereas the AK47 uses a rotating bolt. A simpler trigger mechanism than the AK47, combined with an axial hammer, serve to make this part of the weapon more efficient than its Soviet counterpart.

Supplied in two versions, the Vz58p with either wood or fibre compound stock, fore-end and pistol-grip (earlier versions used wood) and the Vz58v with a folding metal stock (for parachute and armoured troops), the Vz58 can be fitted with a bayonet and flash hider. A bipod is also available but, when fitted, prevents the use of the bayonet.

A tangent leaf rearsight is graduated to 800 m in 100 m increments. The front sight is of the protected post type. Some 1 kg lighter than the AK47, the Vz58 is a well-designed and manufactured weapon.

In service with the Cypriot and Czechoslovak forces.

Vz58

Vz58 with folding stock

M54 sniper rifle

Czechoslovakia

Calibre	7·62 mm × 54 mm R
Dimensions	Length 114·8 cm Barrel length 73·2 cm
Weight	4·1 kg (loaded)
Effective range	800 m
Operation	Bolt
Muzzle velocity	795 m/sec
Sight	Rear—telescopic (× 2½) Front—open adjustable
Magazine	5-round, hinged-box type
Cyclic rate	10 rpm
Manufacturer	State factories

M54 sniper rifle in use

The M54 sniper rifle was developed by the Czechs from the Soviet M1891/30 Moisin Nagant sniper rifle. Although using the same butt and firing mechanism (including magazine) and the same 7·62 × 54R round, the resemblance ends there. The M54 has been considerably reduced in size and weight, but at a cost of a reduced maximum range from 1300 m to only 1000 m. It is, however, claimed to be just as effective at the battle range of 800 m, in spite of being provided with a less powerful telescopic sight. In appearance the main differences between the M54 and the M1891/30 lie in the smaller size of the M54's telescopic sight and the absence of the forward wooden stock and cleaning rod, leaving the M54 with a clean sporting-rifle-type barrel. No bayonet is provided for this rifle, which is likely to be replaced shortly by the Soviet SVD sniper rifle.

In service with the Czechoslovak forces.

Mosin-Nagant rifle

USSR

Calibre	7·62 mm × 54R (Russian rimmed and M1908 (Spitzer))
Dimensions	Length 101·6 cm (132 cm with bayonet extended)
Weight	4·2 kg
Effective range	600 m
Operation	Bolt
Muzzle velocity	820 m/sec
Sight	Front—hooded post Rear—tangent
Magazine	5 round detachable box type
Manufacturer	State factories

(Statistics are for M1944 carbine.)

First adopted by the Russian Army in 1891, the Mosin-Nagant rifle was a combination of a Russian and Belgian design. The weapon is robust but features an unnecessarily complicated bolt, and an awkward safety-catch which requires the cocking-piece to be pulled rearwards and twisted to apply. Produced in enormous numbers, some six variants appeared before production ceased in the 1940s. The main adaptations were in the reduction of length, different sling attachments, and improvements to the iron sights, which, on the early models, did not do justice to the accuracy potential of the rifle. How-

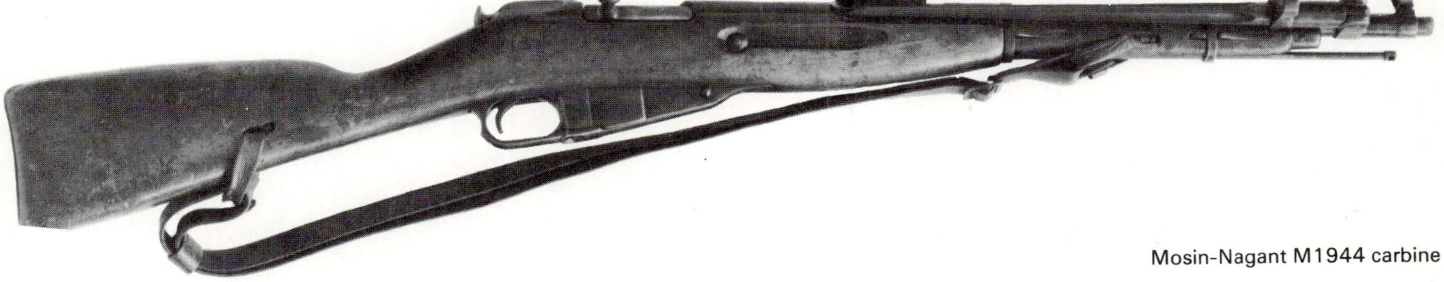

Mosin-Nagant M1944 carbine

ever, when fitted with telescopic sights, the Mosin-Nagant matches up to its other bolt action contemporaries. Models included the M1891/30 rifle and the M1938 carbine. The final model was the M1944 carbine which was fitted with a folding bayonet as opposed to the socket type which all previous models had.

Although now obsolete in the Soviet Army, the Mosin-Nagant is still used by Warsaw Pact countries' second line troops, and was also produced in China as the Type 53 rifle.

In service with some Warsaw Pact second line and other forces in Africa and the Far East.

SKS Simonov carbine

USSR

Calibre	7·62 mm × 39 mm
Dimensions	Length 102·1 cm Barrel length 52·1 cm
Weight	3·8 kg
Effective range	450 m
Operation	Gas semi-automatic
Muzzle velocity	735 m/sec
Sight	Front—hooded post Rear—tangent
Magazine	10 round non-detachable box
Manufacturer	State factories

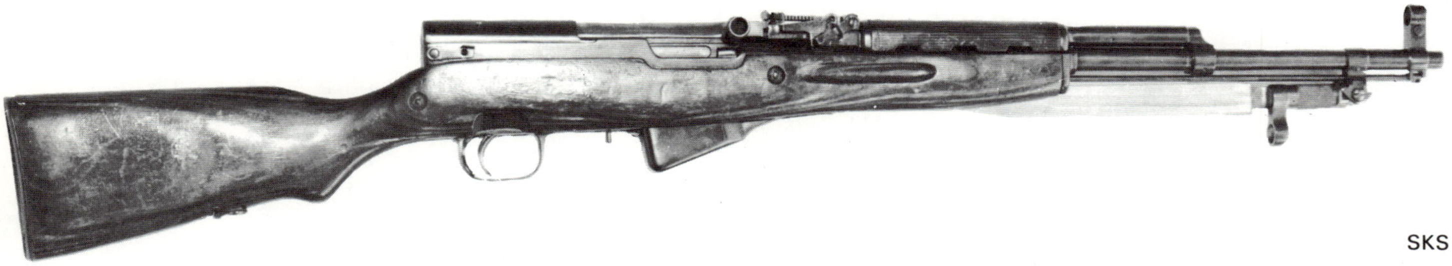

SKS

First designed and produced during the closing stages of World War II, the SKS was the first Soviet weapon to use the now standard 7·62 mm intermediate cartridge.

The action of the weapon is basically a scaled-down copy of the obsolete 14·5 mm PTRS anti-tank rifle. The action is locked by the bolt being tilted into lugs in the receiver by the bolt carrier. The gas system uses a piston with a piston-rod to unlock the action and force it to the rear for the first 20 mm, thereafter inertia takes over combined with the return-spring to complete the loading cycle. The SKS is loaded using either a 10 round charger (called 'stripper clip' in the USA) or individual rounds. The magazine is non-detachable, but the magazine release catch, when actuated, allows the magazine to swing downwards, thus allowing rounds to be spilled out for unloading purposes. Besides this rather unusual feature, the weapon also has a permanently attached folding bayonet which, when not required, lies in a recess in the forestock below the barrel.

Although no longer in first-line service with the Soviet Army (but still widely used for ceremonial purposes), the weapon was produced in vast numbers, and has appeared in slightly modified form in all communist countries. For instance, the Yugoslav version features an integral grenade launcher on the muzzle, and is designated M59. The Chinese Type 56SKS series rifle features a longer barrel and a needle bayonet.

In service with the Angolan, Chinese, North Korean, Vietnamese, Yugoslav and all Warsaw Pact forces.

Kalashnikov AK47 and AKM rifles

USSR

Calibre	7·62 × 39 mm
Dimensions	Length 86·8 cm (AKM—87·6 cm) Barrel length 41·4 cm
Weight	4·3 kg (AKM—3·1 kg)
Effective range	400 m
Operation	Gas selective
Muzzle velocity	710 m/sec (AKM—715 m/sec)
Sight	Front—protected post Rear—tangent
Magazine	30 round detachable box type
Cyclic rate	600 rpm
Manufacturer	State factories

Although, at the end of World War II the Russians had begun introducing the SKS rifle, using the new 7·62 × 39 mm M43 intermediate cartridge, the need was felt for one weapon to replace both the venerable sub machine guns (PPSH41 and PPS43) and the variety of rifles then still in service, using the old M1908 rimmed case. As over 30 million have been produced to date in many countries, albeit with variations, the AK seems to meet the requirement.

First produced within 4 years of the cessation of hostilities, this gas-operated weapon uses a rotating bolt, engaging locking shoulders in the receiver, to lock the action. The bolt carrier and piston rod are attached as one piece. A blank firing device, night sights and a bayonet may be fitted. An old-fashioned, but nevertheless useful, feature is the stowing of a cleaning rod below

AK47 (Polish model)

AK47 with folding stock

AKM

AKM with folding stock

AKM rifle (GDR version). Note the high impact plastic butt and upper forestock attachment. Luminous flip-up night sights are standard equipment

AKM rifle (Rumanian folding stock version). Note the 40 round magazine and forward pistol grip. The muzzle compensator can be unscrewed and replaced by a grenade launcher attachment

the barrel. A folding stock version is also available. A hold-open device and a flash eliminator are not fitted as standard, which is unusual for a weapon of this type.

The AK47 is now being replaced by the AKM. Merely a lightened and modified version, the AKM uses a stamped-steel receiver, a ribbed receiver cover, and a cyclic rate reducer is incorporated in the trigger mechanism. The bayonet, which is not interchangeable with the AK47, can, together with the scabbard, be used as a wirecutter. It is also made with a folding stock. Hungary produces a sub machine gun version of the AKM. East German versions have been fitted recently with a sideways folding frame butt—similar to the Armalite rifle.

Standard in all Warsaw bloc countries and China (Type 56), the Kalashnikov is also produced in Yugoslavia (M64) and Finland (M60 and M62), and it may be encountered in most third world and communist-supported irregular forces.

In service with most communist, Egyptian, Israeli, Syrian and some third world forces.

SVD Dragunov sniper rifle

USSR

Calibre	7·62 mm (M1908), 7·62 mm × 54 mm
Dimensions	Length 122·4 cm Barrel length 61 cm
Weight	4·4 kg
Effective range	900 m
Operation	Gas
Muzzle velocity	830 m/sec
Sight	Front—protected post Rear—tangent
Magazine	10 round detachable box type
Cyclic rate	30 rpm
Manufacturer	State factories

The Dragunov is a semi-automatic sniping rifle and, like the French FR-F1, was designed for that purpose.

Using the basic and well-proven Kalashnikov action, the weapon has been strengthened to handle the old M1908 7·62 mm rimmed cartridge. The trigger mechanism has been modified to give a more positive pull-off. Issued with the PS01 telescopic sight, the rifle has a cheek-piece attached to the rather unusual skeleton stock with pistol-grip. Ordinary iron tangent sights are also fitted. The muzzle is fitted with a combined flash eliminator/muzzle brake, and a bayonet lug for the standard AKM type bayonet.

The Dragunov is now replacing the M1891/30 bolt-action rifles, although large numbers of these will still be found in use by satellite countries.

In service with the Soviet and other Warsaw Pact forces.

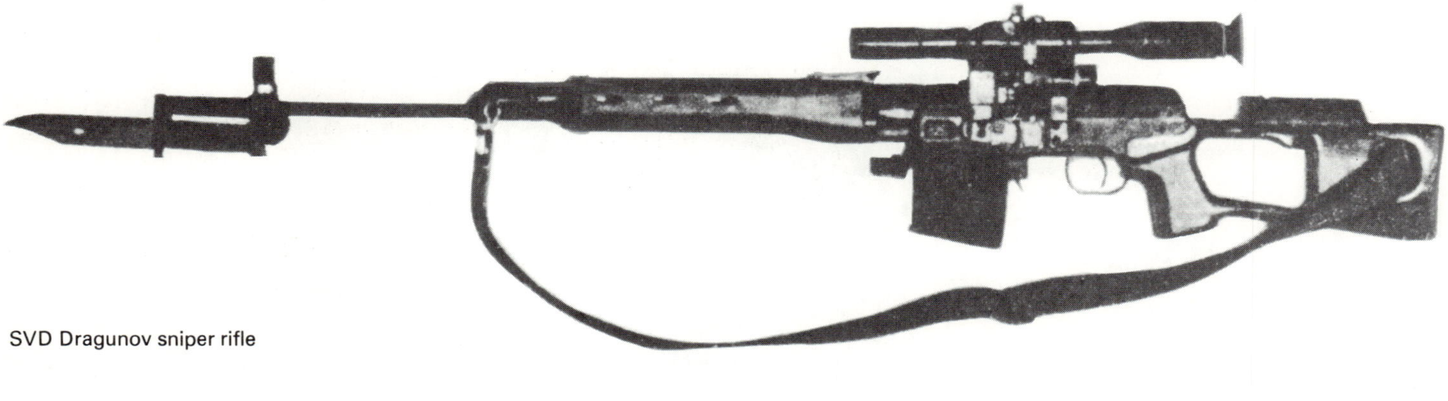

SVD Dragunov sniper rifle

M59/66 rifle

Yugoslavia

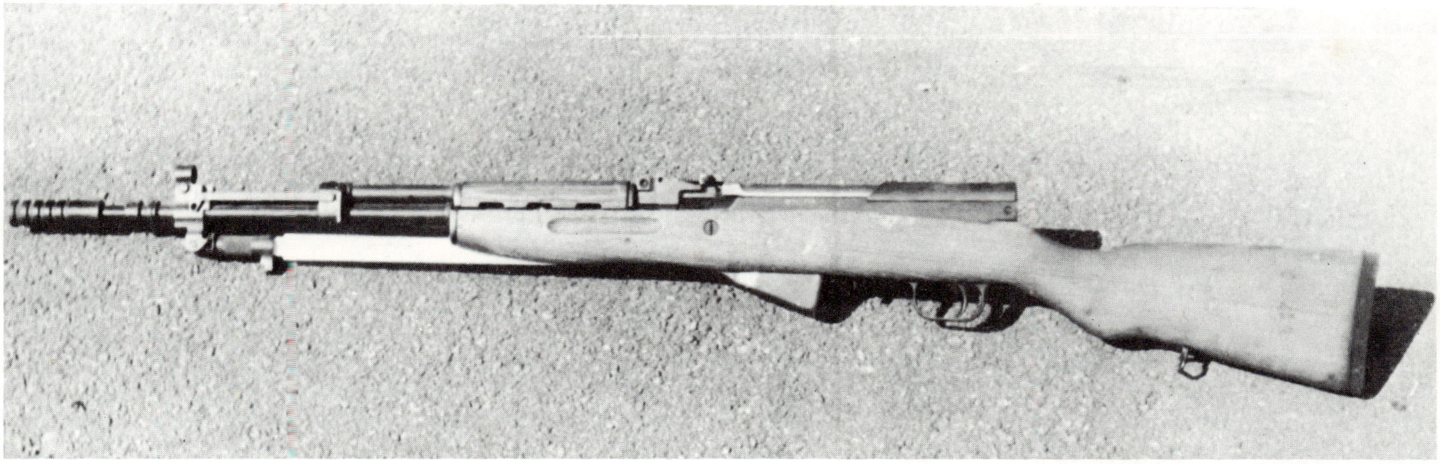

(above) M59/66 carbine, with rubber butt pad

M59/66 carbine grenade launcher sights and luminous foresight for night operations. The range gradation on the left—up to 240 m for the M60 AP rifle grenade and, right, up to 150 m for the M60 HEAT grenade

Calibre 7·62 mm × 39 mm
Weight 4·2 kg
Manufacturer State factories

The M59/66 rifle is a modification of the M59, the Yugoslav copy of the Soviet SKS, and has replaced it as the standard Yugoslav army rifle. The modifications consist of a permanent fixture of a spigot-type grenade launcher and grenade launcher sight, the latter being hinged behind the normal foresight, and an increase in the length of the folding knife bayonet by 6 cm. It can be used as a normal rifle when the grenade launcher sight is folded down on the top of the barrel. See the SKS for further details.

In service with the Yugoslav forces.

M64 rifle

Yugoslavia

Calibre 7·62 × 39 mm
Dimensions Length 107 cm
Barrel length 51 cm
Weight 3·8 kg
Operation Gas—selective. Fires M60 series grenades
Magazine 20, 30 and 40 round box magazines
Manufacturer State factories

There are three versions of this weapon, which is the Yugoslav version of the AK47, the long barrel, the short barrel and the folding stock types. The most important feature of this weapon is that it has a hold open device, indicating to the firer when the magazine is empty. Originally developed with a 20 round magazine, 30 and 40 round magazines are currently being issued.

M64

M64 rifle

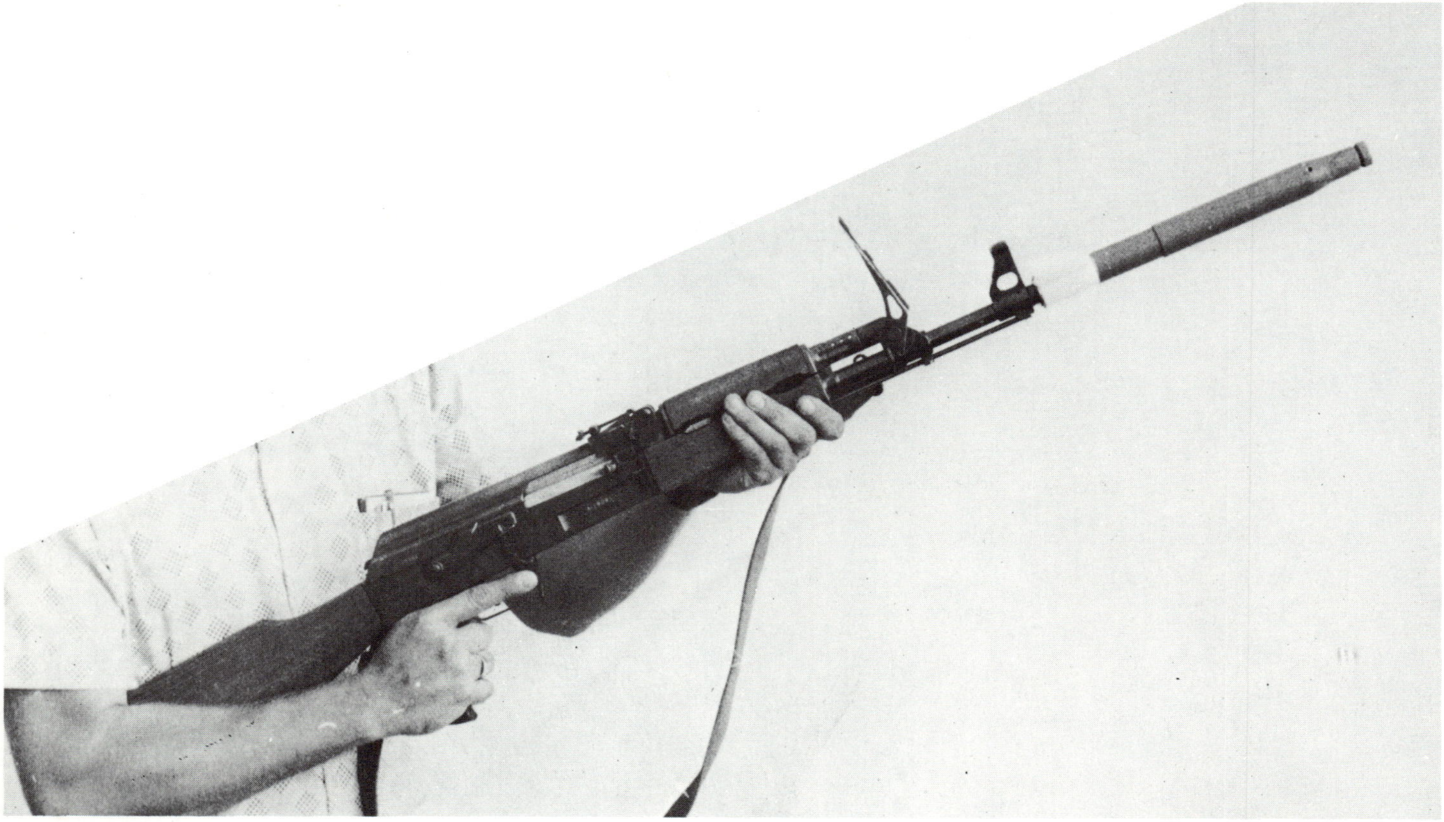

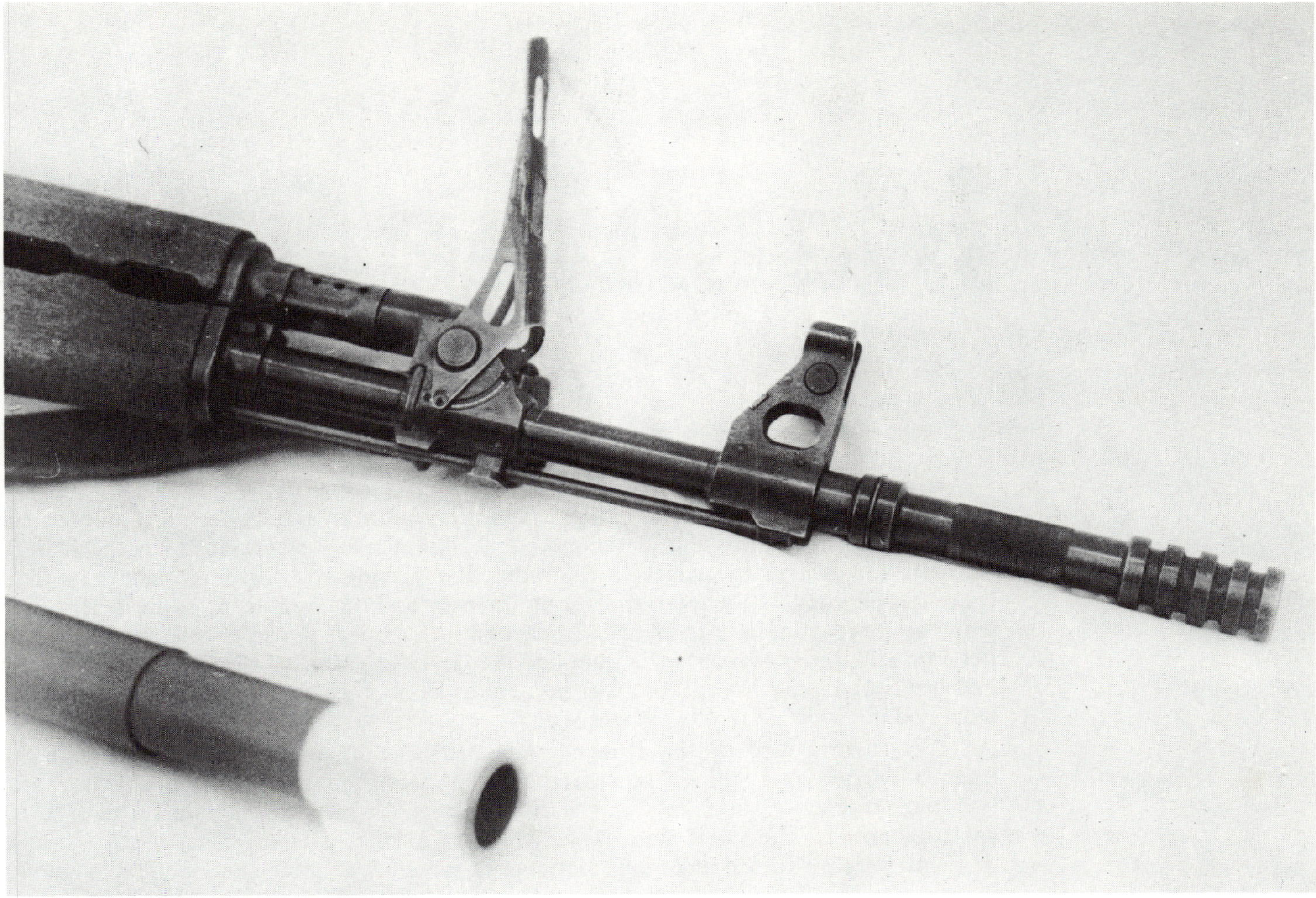

M64 with grenade launcher attachment and sights

M64 with M60 HEAT grenade

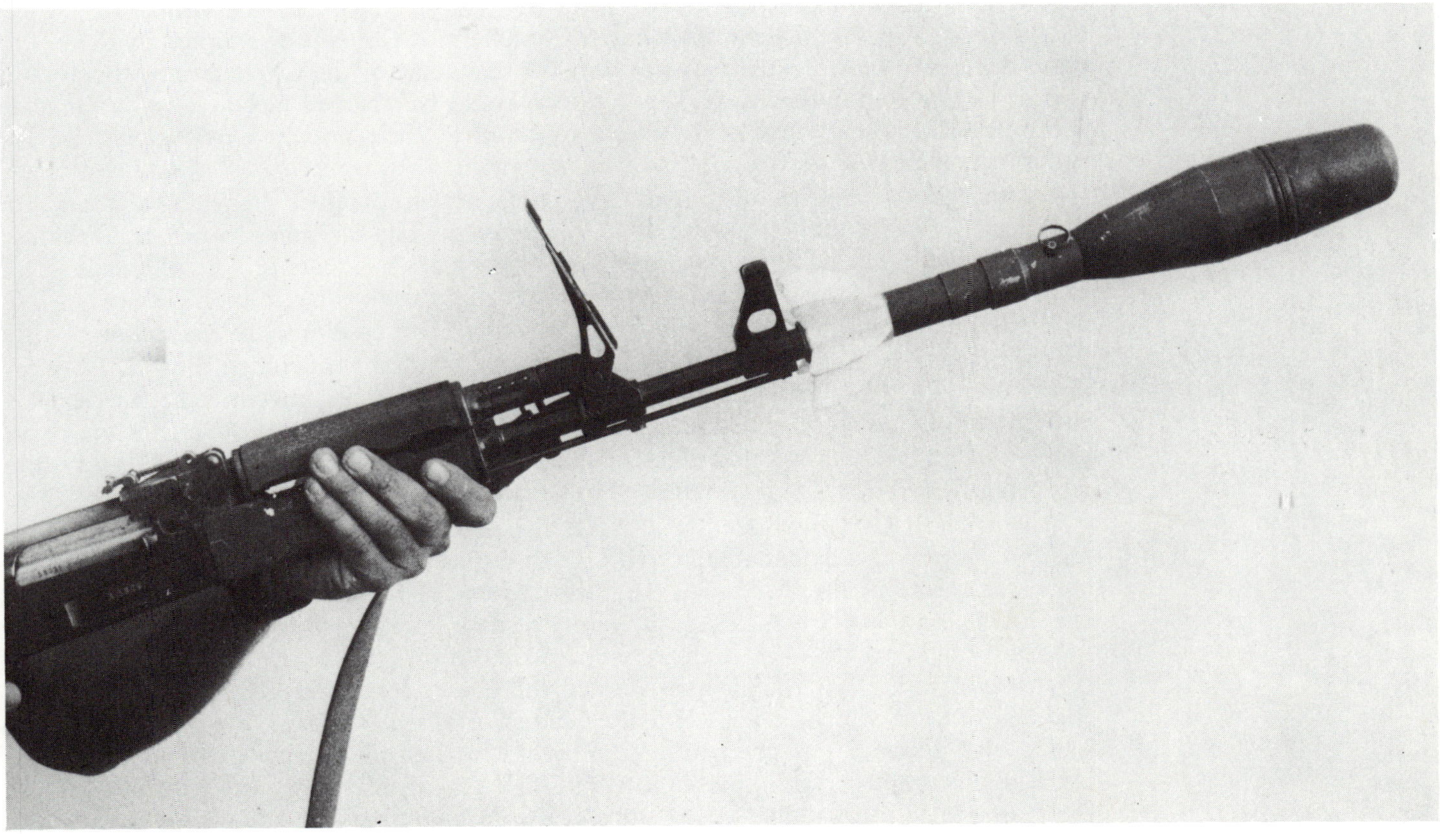

Machine Guns

Prior to World War I, the machine gun was not readily thought of as an infantry close support weapon, but was considered more as a light artillery piece. However, the British, German and American armies were gradually replacing the wheeled mounts of their water-cooled guns with tripods for use by infantry and the Americans had experimented with the Lewis gun, despite an initial rejection, after it was put into series production in Belgium and later produced in Britain and the USA. The Madsen LMG, which the Danes had been producing since 1904, was becoming popular with many smaller nations and remained in service until after World War II.

During World War I, the British made extensive use of the Lewis gun, but the Americans, for some still unknown reason, were issued mostly with the awkward and unreliable French Chauchat. Their native-designed Browning Automatic Rifle (BAR) appeared only for the 3 or 4 months preceding the Armistice in November 1918. All these weapons were air-cooled and could not give sustained fire due to the inability to change the barrel. The Germans, however, preferred to adapt their proven 08 Maxim water-cooled gun into a light machine gun by adding a buttstock, pistol-grip and bipod, thus making the weapon (MG 08–15) capable of being used by one man, so long as the belt was contained in the box which could be attached to the side of the gun.

Shortly after the end of World War I, the French, greatly impressed with the American BAR, went on to design their BAR-based Chatellerault which they adopted in 1924. The main difference between the two was that the magazine of the Chatellerault was top- as opposed to bottom-mounted, and the weapon could be stripped more easily by hingeing the buttstock. The British and Americans retained their proven Lewis guns, and the Americans their BAR as well.

These three countries also maintained production of their heavier weapons, the Hotchkiss, Vickers and Browning 1917 respectively. The Germans, however, forbidden the use of water-cooled machine guns by the articles of the Treaty of Versailles, converted their earlier Dreyse water-cooled guns to air-cooled for use by the infantry.

At that time, military opinion held that sustained fire coud only be maintained by the use of a water-cooling jacket, and so the League of Nations concluded that the Germans were then effectively denied a sustained fire capacity. To circumvent this, the Germans proceeded, not only to make the gun air-cooled, but also to modify it so that the barrel could be rapidly detached and changed, thereby eliminating the need for water-cooling. Known as the MG13, the gun remained in service as standard until 1935.

During the 1920s, the Czech firm of ZB at BRNO were working on a design of a light machine gun with a detachable barrel. Introduced as the ZB26, in 7·92 mm, the weapon later went on to be produced with modifications by the British from 1938 in ·303 in and was known as the Bren. Although a variety of mounts was available for the gun during the course of World War II, it never replaced the Vickers gun in the sustained fire role. It is worthy of note that the British adopted the Czech design in preference to their own Vickers-Berthier which, along with the Danish Madsen, underwent extensive trials in 1932. This decision was one of the first breaks in the then somewhat nationalistic and traditionalist military thinking in Britain. The Americans, meanwhile, had developed their air-cooled Browning M1919 tank gun into an infantry model on a tripod (later superseded by the light M2 tripod) to become the M1919A4. This model remained standard until well after the Korean War. A LMG version, known as the M1919A6, was fitted with a buttstock, pistol-grip and bipod but, due to its lack of attachable belt container, remained as a 2-man operated weapon. The BAR was also retained until well after Korea.

During this period most large countries with their own arms plants were also turning out LMGs, many of them based on weapons already proven in service with other armies.

Gas operation was becoming popular for the LMG and, although belt and clip feed were still prevalent, the value of a relatively dirt-free and durable box magazine was being realised. The drum or pan magazine, and particularly that of the Lewis, caused stoppages if even slightly bent.

Refining on the idea of the MG13, the Germans began series production of their MG34 in 1936. What might be classed as the first true general purpose machine gun (GPMG), the MG34 was made available with a wide variety of mounts ranging from a buffer-mounted tripod which could be fitted with telescopic sights to AA or heavy pillbox mounts. Although belt-fed, the gun could easily be carried and fired by one man by inserting a 50 round belt into a round basket drum which was then attached to the feed-tray. By removing the top cover and fitting an adaptor, the 75 round saddle drum (originally produced for use with the MG15 aircraft gun) could be utilised. This spring-loaded magazine raised the rate of fire and, used in conjunction with a large ring sight, made the weapon well suited for anti-aircraft use. With the addition of an armoured barrel-jacket and removal of the select fire mechanism, the gun became the standard tank co-axial machine gun. Although not renowned for its reliability under adverse conditions due to the close machining tolerances, it did set the precedent for this type of multi-purpose weapon.

During World War II, the MG34 was superseded by the MG42. Cheaper to produce due to the wide use of stampings, the MG42 was lighter and had a higher rate of fire than its predecessor. The gun could be used with almost as wide a variety of mounts and was considerably more reliable.

Among the numerous designs to come out of Germany during the course of the war was the FG42. The Luftwaffe, ever aware of the specialist role of their paratroops, first gave them the MP38 SMG before the war. (This weapon was originally issued to armoured troops as well and went on to be used by men of all arms.) Later, as the need was felt for a lighter support weapon than the MG34, the FG42 appeared. Again made primarily from stampings, this gas-operated weapon featured a hitherto unique fire selective device. When firing single shots, the weapon fired from a closed bolt for greater accuracy and from an open bolt on automatic to give greater cooling capacity. This system has since been used on several post-war weapons, although the wartime US Johnson MG also featured this device. Only approximately 5000 FG42s were manufactured, but the weapon was popular and is reported to have been reliable and accurate.

At the end of World War II, most countries continued to use their wartime weapons, precedent being put on rebuilding the economies rather than re-equipping the armies. However, as the formation of NATO and other defensive alliances to meet the threats took place, so the need to update the member nations' armies became urgent.

NATO, having finally decided to standardise on the 7·62 × 51 mm round, then proceeded to have various countries tender designs for weapons to take this cartridge.

The Belgians, admirably aware that a gun designed using the best features of well-proven weapons can often be preferable to something entirely new, proceeded to produce the MAG GPMG. Copying the feed mechanism of the MG42 and the bolt of the BAR, the MAG, though expensive to manufacture, has been supplied in modified form to over 20 countries. The gun, fitted with a telescopic or infra-red night sight, may be used on a bipod or tripod. Mounts are also available for fitting on to soft-skinned or armoured vehicles and single or dual mounts are fitted to helicopters on pods or brackets.

The Germans, content that their MG42 has withstood the test of time, continue to use the weapon in 7·62 mm with some modifications. The gun is also manufactured under licence in Italy, where it is sometimes fitted with a heavier bolt to reduce the rate of fire. Yugoslavia still produces the weapon in its original 7·92 mm calibre as the SARAC.

The Americans, like the British before the war, were prepared to concede to proven designs in preference to native ones and so the 7·62 mm US M60 machine gun, like the MAG, used the MG42 feed mechanism and the bolt of the FG42.

In the Warsaw Pact countries, the 7·62 mm M43 cartridge was designated the official calibre, and the adoption of this intermediate cartridge led to a series of LMGs, the main designs being produced by the Soviet and Czech arsenals. The Czechs, particularly, produced some interesting weapons, the most revolutionary being the Vz52 series. Not only capable of taking either belt or magazine feed without the addition or subtraction of any parts, the weapon cocked by withdrawing the pistol-grip assembly in much the same way as the British BESA tank gun.

The earlier Soviet RP46 and RPD LMGs (merely developments of the old Degtyarev action) were later replaced by the RPK, which was basically a LMG version of the Kalashnikov rifle, and the two weapons shared a fair degree of parts interchangeability.

The Soviet Union, although a great believer in the intermediate cartridge, has retained its old M1908 7·62 mm rimmed cartridge for use in their PK GPMG. Again using the Kalashnikov action, the PK may be fitted with a heavier barrel for use in tanks and in the sustained fire role.

Today, some countries, in addition to the LMG or GPMG, produce a squad automatic rifle (SAR) version of their assault rifle. Fitted with a bipod and heavier barrel, the SAR serves as an extra section support weapon. The SAR versions of the Italian BM59 and the Belgian FAL are examples of this.

ZB26/30 light machine gun

Czechoslovakia

Calibre	7·92 mm × 57 mm, 7·65 mm × 54 mm ·303 in, ·30–06, 7 mm
Dimensions	Length 116·5 cm Barrel length 67 cm
Weight	9·6 kg
Effective range	900 m
Operation	Gas
Type of fire	Selective
Muzzle velocity	760 m/sec (in 7·92 mm)
Sight	Front—blade Rear—drum aperture
Feed system	20 round detachable box magazine
Cyclic rate	500 rpm
Barrel change	Yes
Manufacturer	State factories

Although no longer in front line service in any major army, the ZB26, ZB30 and derivatives were produced by, or sold to, so many nations in such numbers and calibres that the weapon will certainly be encountered for some years to come. Originally designed in 1924 by Vaclav Holek, the gun was slightly modified to become ZB30. Both models have been manufactured in China, the ZB30 in Iran and Rumania, and in Yugoslavia as the Model 30J. The ZB is the forerunner of the British Bren gun, and was produced in 7·92 mm (standard), 7·65 mm, 7·00 mm, ·303 in and ·30 in. Spain produced the weapon in 7·62 mm (NATO) with belt feed called the FAO.

The action is locked using a simple tilting bolt engaging in recesses in the top of the receiver. The barrel is heavily finned for cooling purposes and is quickly detachable using a rotating-barrel locking-nut. The gun may be used on a tripod, or with the standard bipod which is adjustable for height. A hinged butt-plate, when not in use, lies along the top of the stock. Although normally used with a 20 round magazine, locally produced extended magazines have been encountered. An expensive drum-type rear sight is fitted. Due to the large amount of machining required to produce the weapon, and particularly the long receiver/piston housing and finned barrel, the gun was not cheap. Nevertheless, it functions very reliably and is renowned for its accuracy.

In service with reserve and guerrilla forces worldwide.

ZB30 LMG

Vz52 and Vz52/57 light machine guns

Czechoslovakia

Calibre	Vz52—7·62 mm × 45 mm (M52 Czech) Vz52/57—7·62 mm (M43), 7·62 mm × 39 mm
Dimensions	Length 104 cm Barrel length 54·5 cm
Weight	8·7 kg
Effective range	800 m
Operation	Gas
Type of fire	Selective
Muzzle velocity	745 m/sec (M52)
Sight	Front—blade with removable hood Rear—U notch
Feed system	25 round box magazine or 100 round belt
Cyclic rate	900–1000 rpm
Manufacturer	State factories

Although outwardly resembling a normal LMG, the Czech Vz52 and Vz52/57 differ in many respects from other more conventional types. The locking system of this gas-operated gun is on the same tilting bolt system as its forerunner, the ZB26, which, with modifications, became the Bren gun. However, this gun is unusual in that a box magazine or a belt may be used without the addition of any extra parts. By merely closing the belt feed cover and opening the top cover, a box magazine may be fitted, and vice versa. The cocking-handle is dispensed with, and the action is retarded by pulling the pistol-grip/trigger mechanism to the rear. The type of fire is selected by pulling the top part of the trigger for single shots and the lower part for fully automatic, in the same way as the German MG34.

The barrel, which is fitted with a variable gas cylinder similar to the Bren, may be removed by rotating the top-feed cover to the right and pulling the barrel forward.

The original Vz52 was produced in 7·62 mm × 45 mm (M52 Czech) cartridge. However, the gun was later changed to accept the Soviet 7·62 mm (M43), thus becoming the Vz52/57. The Vz52 is now obsolete.

In service with the Czechoslovak forces (Vz52/57).

Vz52 LMG

Vz52 LMG, showing 100 round belt, magazine and spare barrel

Vz59 general purpose machine gun

Czechoslovakia

Calibre	7·62 mm M1908 and 7·62 mm NATO
Dimensions	Length 121 cm (Vz59), 111 cm (Vz59L) Barrel length 69 cm (Vz59), 59 cm (Vz59L)
Weight	19 kg (Vz59 with tripod), 8·6 kg (Vz59L)
Effective range	800 m
Operation	Gas
Type of fire	Automatic
Muzzle velocity	830 m/sec
Sight	Front—blade with hood Rear—leaf
Feed system	50 or 250 round link belt
Cyclic rate	700–800 rpm
Barrel change	Yes
Manufacturer	State factories

The general purpose Vz59 machine gun is now the standard machine gun of the Czech Army and, like the Soviet PK, uses the old M1908 7·62/54 mm rimmed cartridge. An export version in 7·62 mm (NATO), the Vz59N has been purchased by a few small nations.

The gun is capable of several variations. When mounted on its tripod (which can be opened up to elevate the gun for anti-aircraft use), it is known as the Vz59. When designated Vz59L, the gun is fitted either with a light barrel and detachable bipod as a section automatic rifle, or with a heavier and longer barrel and tripod as a light machine gun. The solenoid-fired version for tank use is the Vz59T.

The weapon utilises the same sliding-trigger cocking mechanism and quick-change barrel of its forerunner, the Vz52. However, the old dual-feed belt and box magazine have been replaced with belt feed only, from a 250 round belt or a 50 round container which can be mounted on the gun. The swinging lug lock is very similar to that used in the Vz58 assault rifle. Gas flow may be adjusted using the two-position type of regulator. On the 7·62 mm (NATO) version, a four-position regulator is fitted, giving a rate of fire exceeding 1000 rpm for anti-aircraft use when used in the number 4 position. A hinged butt-plate is standard and, apart from the normal sights, infra-red sights or a ×4 telescopic sight may be attached to any version.

In service with the Czechoslovak forces.

Vz59 GPMG

Vz43 heavy machine gun

Czechoslovakia

Manufacturer	State factories

See the SG43 and SGM.

In service with the Czechoslovak forces.

1MG-K light machine gun

Germany (DGR)

Manufacturer State factories

See the RPK.

In service with the DGR forces.

Wz43 heavy machine gun

Poland

Manufacturer State factories

See the SG43 and SGM.

In service with the Polish forces.

SB30 light machine gun

Romania

Manufacturer State factories and in Iran

See the ZB26/30.

In service with the Iranian and Romanian reserve forces.

DP and DPM light machine guns

USSR

Calibre	7·62 mm M1908, 7·62 mm × 54 mm
Dimensions	Length 127 cm Barrel length 60 cm
Weight	9 kg
Effective range	800 m
Operation	Gas
Type of fire	Automatic
Muzzle velocity	840 m/sec
Sight	Front—protected post Rear—tangent
Feed system	47 round pan
Cyclic rate	600 rpm
Manufacturer	State factories

DPM LMG (Chinese version)

RP46 light machine gun

USSR

Calibre	7·62 mm M1908, 7·62 mm × 54 mm	*Type of fire*	Automatic
Dimensions	Length 128 cm Barrel length 61 cm	*Muzzle velocity*	840 m/sec
		Sight	Front—protected post Rear—tangent
Weight	13 kg	*Feed system*	250 round belt or 47 round pan
Effective range	800 m	*Cyclic rate*	600 rpm
Operation	Gas	*Manufacturer*	State factories

RPD light machine gun

USSR

Calibre	7·62 mm × 39 mm	*Muzzle velocity*	740 m/sec
Dimensions	Length 103 cm Barrel length 52 cm	*Sight*	Front—protected post Rear—tangent
Weight	7 kg	*Feed system*	100 round belt
Effective range	800 m	*Cyclic rate*	650–750 rpm
Operation	Gas	*Manufacturer*	State factories
Type of fire	Automatic		

RP46 LMG

RPD LMG

The DP first appeared in 1926, and was later developed to give many different configurations of the basic weapon.

A gas-operated weapon, the action is locked using flaps cammed by the firing pin into recesses in the receiver. A weak feature of the early system was the return spring. Coiled around the piston, it weakened and distorted easily due to barrel heat. The DP used a 47 round pan-type magazine. A tank version known as DT was also produced using a variable-length stock, a 60 round magazine, a pistol-grip and a non-detachable barrel.

During World War II both versions were modified by placing the return spring in a cylindrical container behind the action. The addition of a pistol-grip and a stronger bipod completed the modification and the weapons were then known as the DPM and, as a tank version, the DTM. Using the same action the RP46 is a basic variant of the DP and was designed as a company support weapon. The weapon is belt-fed, and a carrying handle has been added over the forward end of the body, but not to the heavier duty barrel which, like that of the DP and DPM, is detachable. The DP 47 round magazine may also be used alternatively by removing the belt-feed device, to make the gun more versatile.

The final version in the Degtyarev range is the RPD. Firing the 7·62 mm M43 intermediate cartridge, the weapon uses the same scaled-down type of gas and bolt-locking system as the DP series and the RP46. A purely belt-fed gun, the belt was contained in a drum which was attached below the gun. This allowed for easier carriage and use by one man if necessary. A variant is the RPDM which uses a longer piston.

Although now obsolete in the Soviet Army, all these weapons are still to be found in use in Warsaw Pact and in North Korean (Type 62 LMG) forces. The communist Chinese have also produced almost exact copies of these weapons (Types 56 and 56–1 LMG).

In service with the Chinese, Egyptian, North Korean, Pakistani and Vietnamese forces.

RPK light machine gun

USSR

Calibre	7·62 mm M43, 7·62 mm × 39 mm
Dimensions	Length 103 cm Barrel length 61 cm
Weight	5 kg
Effective range	800 m
Operation	Gas
Type of fire	Selective
Muzzle velocity	735 m/sec
Sight	Front—protected post Rear-tangent
Feed system	30 or 40 round box magazines, or 75 round drum magazine
Cyclic rate	600 rpm
Manufacturer	State factories

This gas-operated LMG is a development of the AKM assault rifle and has replaced the old RPD as the section support weapon.

The weapon differs from the AKM in several respects. A longer and heavier barrel with a folding bipod is fitted, though there is no provision for the fitting of a bayonet. The barrel is not detachable. The weapon may feed from a 40 round box magazine, a 75 round drum magazine or the standard AK 47 or AKM 30 round magazine. Any of these magazines may be used with any Kalashnikov weapon. All parts are interchangeable with the AKM except the front hand guard, buttstock, rearsight and the modified receiver. Some RPKs are capable of mounting an infra-red sighting device. The weapon is also available in Yugoslavia (M65A and B) and in East Germany (1MG-K).

In service with some Arab, the Soviet and other Warsaw Pact and Yugoslav forces.

RPK LMG (showing drum and box type magazines)

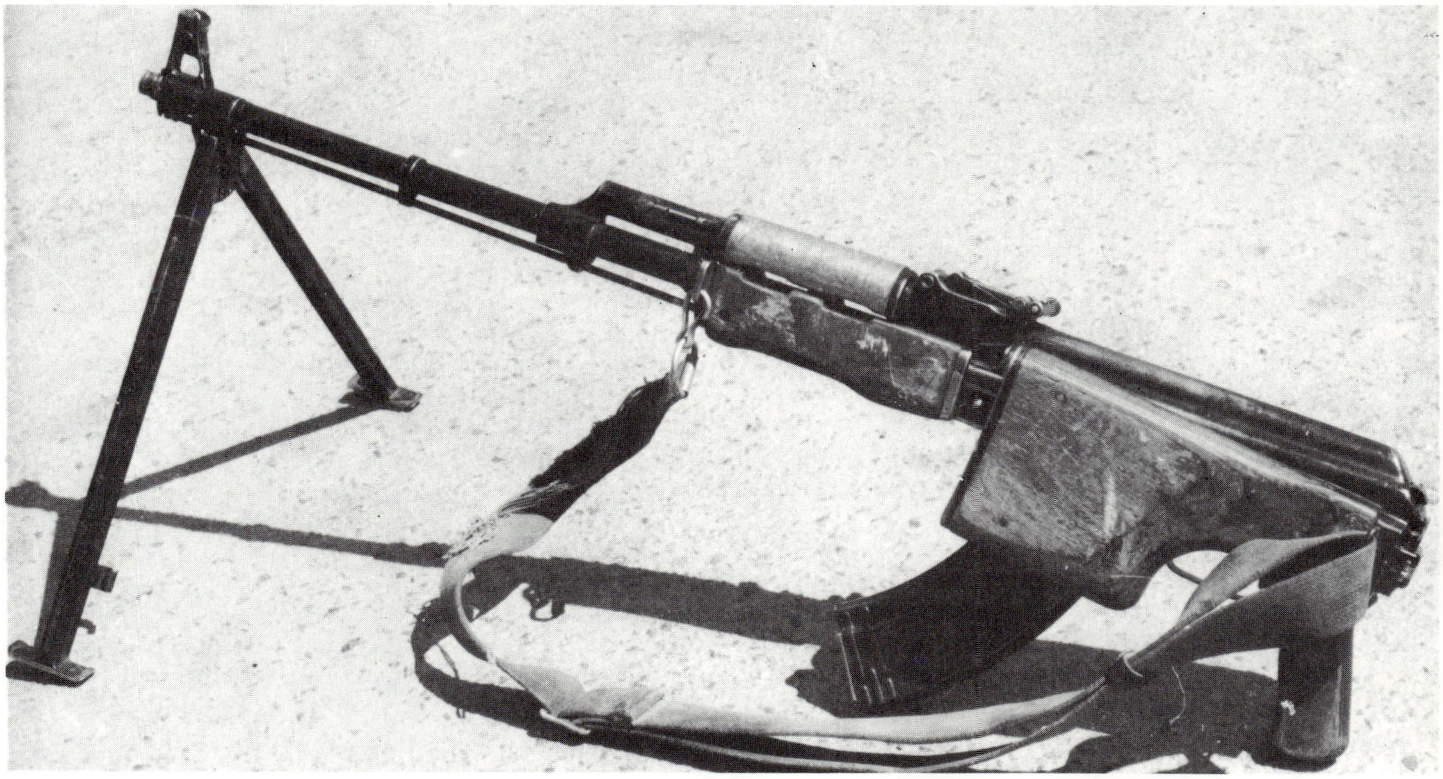

RPKS with a folding buttstock; otherwise identical with the RPK

Close-up of the RPKS, showing the hinge

PK general purpose machine gun

USSR

Calibre	7·62 mm M1908, 7·62 mm × 54 mm	*Weight*	8·9 kg
Dimensions	Length 116 cm	*Effective range*	1000 m
	Barrel length 65·5 cm	*Operation*	Gas
		Type of fire	Automatic

Muzzle velocity	825 m/sec
Sight	Front—post with ears
	Rear—tangent
Feed system	50, 200 or 250 round link belt
Cyclic rate	650 rpm
Barrel change	Yes
Manufacturer	State factories

The PK is the latest machine gun in use with the Warsaw Pact countries. A general purpose machine gun, the weapon may be used in three roles. The LMG version known as the PK is used with the adjustable bipod and fed from a choice of 50, 200 or 250 round square belt boxes. The 50 round box may be attached to the bottom of the gun in the same way as the RPD. When used as a heavy machine gun the weapon is designated PKS and uses a tripod. An extension for the tripod is available making the gun suitable for use in the AA role. Equipped with a solenoid firing device, the gun becomes the PKT for use in armoured vehicles. A gas regulator and carrying handle is fitted to the quick change barrel and a bar-type flash suppressor has replaced the earlier cone-type flash hider.

Being a Kalashnikov design, the action is locked using the familiar turning bolt system. The feed mechanism is operated by two cams on either side of the one-piece piston/bolt carrier which cause the vertically mounted feed lever to pivot inwards on recoil and outwards on counter-recoil. The round is extracted from the belt by a cartridge gripper mounted on the carrier. When the carrier is driven rearward, the round is forced on to the feed tips by a depressor working in conjunction with the feed-cam. The PK uses a light skeleton stock and some models are fitted with brackets for infra-red sights.

In service with the Warsaw Pact forces.

PK GPMG

PKM GPMG

PKM light machine gun

USSR

The PKM differs from the more common PK model in that the barrel flutes between the carrying handle and gas port have been removed and the muzzle cone has been replaced by an FN-type flash eliminator. The PKM is several kilos lighter than the PK, the reduction to 8 kg being made by replacing the machined parts by pressings and stampings on the body assembly. The performance characteristics remain the same.

PKM-LMG

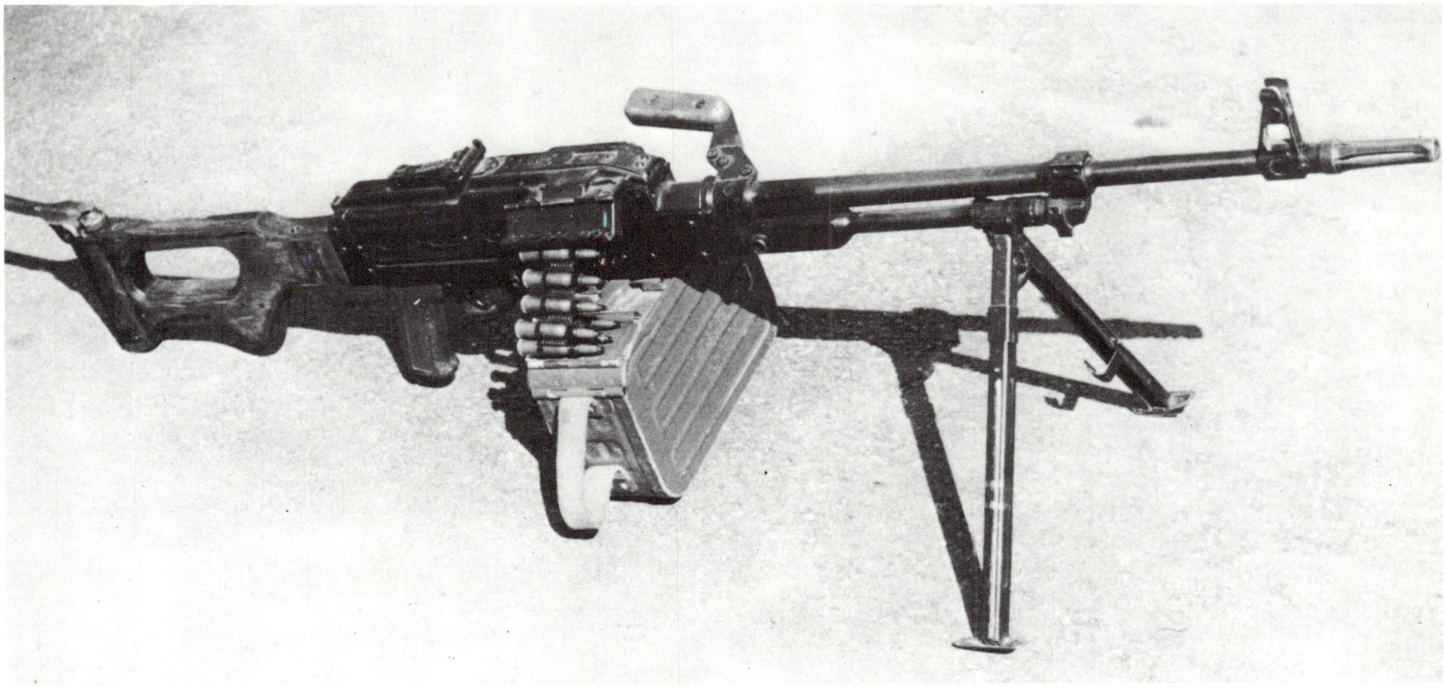

Close-up of the PKM LMG, showing the feed aperture for the belt contained in the 100 round detachable liner

Goryonov SG43 and SGM medium machine guns

USSR

Goryonov SG43 MMG in the anti-aircraft role

Goryonov SG43 MMG

Calibre	7·62 mm M1908, 7·62 mm × 54 mm
Dimensions	Length 112 cm Barrel length 71·5 cm
Weight	13·5 kg
Effective range	1000 m
Operation	Gas
Type of fire	Automatic
Muzzle velocity	745 m/sec
Sight	Front–blade Rear—leaf suitable for two differently powered loads
Feed system	250 round link belt
Cyclic rate	600–700 rpm
Barrel change	Yes
Manufacturer	State factories

The Goryonov series of medium machine guns is used by almost all of the European and Asian communist countries, and has been manufactured in China and Hungary. The Soviet-produced weapon comes in two basic versions, the SG43 and SGM. Slightly modified versions of both models have been produced, though these differ only in minor details, such as the addition of dust covers and different barrel locks. The main recognisable features are the smooth barrel with the cocking handle mounted between the spade grips, containing oiling brushes, of the SG43, as opposed to the splined barrel and right-side mounted cocking handle of the later SGM. Both versions may be used on a Sidorenko-Malinovsky tripod, or on a wheeled mount with an optically mounted shield. The SGM was modified for use with the older Sokolov-type wheeled mount which was used with the now obsolete Maxim water-cooled machine guns. Both wheeled mounts may be upturned and used as anti-aircraft mounts. Vehicle mounts are also available (SGMB), and a solenoid-fired version of the gun, known as the SGMT, was produced for use in armoured vehicles. All models have a quick-change barrel which has a chrome-lined bore and is fitted with a carrying handle. The action is locked using a tilting bolt which is unusual in that it recesses into the right-hand side of the receiver. The cartridge-gripper is held by a cartridge feed slide which, in turn, is attached to the bolt. As the bolt is withdrawn, the gripper extracts the cartridge from the belt and, having cleared the feed tray, is forced downwards by the cartridge-depressor to be fed into the chamber by the forward motion of the bolt. It is interesting to note that the cartridge feed slide may be quickly removed without stripping the weapon, and many of these guns captured in Vietnam were found to be totally inoperable due to this vital component having been jettisoned by the Viet Cong.

The Goryonov weapons are being replaced by the PK general purpose machine gun. The Chinese version of the SG43 is known as the Type 53 heavy machine gun and the SGMB as the Type 57 heavy machine gun. The Czechoslovak and Polish manufactured versions of the SG43 are stamped Vz43 and Wz43 respectively. A version produced in Hungary uses the same action, but is fitted with a pistol-grip trigger mechanism and RPD-type buttstock and bipod. The Hungarian gun is used as a GPMG in the same way as the PK. When firing the Type L or LPS light bullets, the range figures on the right-hand side of the leaf-sight should be used and the left-hand side figures with the yellow-tipped heavy bullets.

In service with the Egyptian, Palestinian and Vietnamese forces.

DShK 38/46 heavy machine gun

USSR

Calibre	12·7 mm × 108 mm	*Sight*	Front—post with ears Rear—leaf
Dimensions	Length 158·5 cm Barrel length 107 mm	*Feed system*	50 round link belt
Weight	35·5 kg	*Cyclic rate*	540–600 rpm
Effective range	2000 m 1000 m (anti-aircraft)	*Barrel change*	Yes
Operation	Gas	*Manufacturer*	State factories
Type of fire	Automatic		
Muzzle velocity	860 m/sec		

The DShK 38/46 is the standard heavy machine gun of most European and Asian communist countries,

DShK 38/46 HMG

DShK 38/46 HMG in close-up

although the latter may still use the original M38. The weapon may be mounted either on a vehicle, or on a wheeled mount which may be converted for use in the anti-aircraft role by removing the wheels and unfolding the legs. A special anti-aircraft ring sight is available for use in this configuration and the weapon is fitted in the Soviet standard tank-mounted AA machine gun. The action of the weapon is of the Degtyarev type which is locked, using flaps cammed by the carrier, into recesses in the receiver. The feed mechanism of the M1938 is of the rotating block type with a rounded feed cover, but that of the DShK 38/46 is of the normal shuttle type, and may be adapted to left- or right-hand feed. The barrel is detachable and is fitted with a unique muzzle brake not present on the earlier model. A quadruple mounted version of the gun is produced by the Czechs as a light, wheeled anti-aircraft weapon against targets travelling at up to 930 kph.

The Chinese version DShK 38/46 is known as the type 54 heavy machine gun. Both this and the Soviet version are used by guerillas in the Middle and Far East to defend their bases.

In service with the Cypriot, Israeli, Soviet, other Warsaw Pact and most other communist and some guerilla forces.

KPV Vladimirov heavy machine gun

USSR

Calibre	14·5 mm × 114 mm
Dimensions	Length 200 cm Barrel length 135 cm
Weight	51 kg
Effective range	2500 m, 1500 m (AA)
Operation	Recoil
Muzzle velocity	990 m/sec
Feed system	100 round metal link belt
Cyclic rate	600 rpm
Manufacturer	State factories

This is a modern automatic weapon well suited to both ground and anti-aircraft use. It was designed in the mid-1950s to use the ammunition developed for the anti-tank rifles of World War II, and this gives it a formidable punch when attacking armour. The gun is unusual among Soviet machine guns in that it locks the bolt by turning it and engaging lugs on the bolt-head with the sides of the breech. The recoil operation is also a departure from the normal gas system but has obviously been designed to allow a rapid and easy barrel-change.

The KPV has been made in large numbers and is well distributed among the Warsaw Pact armies. In its AA form it is paired or quadrupled and is then known as the ZPU–2 or ZPU–4. The lighter version is the ZPU–2 which was encountered in quantity in Vietnam, where it claimed a substantial number of US aircraft of all types. Against aircraft the effective range is 1500 m, which makes it formidable. The more usual manner of use now is to mount the gun on Soviet armoured personnel carriers for use as light anti-armour guns, as it is being phased out in the AA role.

In service with the Soviet, Warsaw Pact and communist-supplied forces.

KPV HMG, the ZPV-1 single barrel AA gun

KPV HMG. Close-up of the ZPV-1 feed mechanism

SARAC light machine gun

Yugoslavia

Calibre	7·92 mm × 57 mm
Dimensions	Length 123 cm Barrel length 56·6 cm
Weight	11·6 kg
Effective range	800 m (bipod)
Operation	Recoil, gas-boosted
Type of fire	Automatic
Muzzle velocity	820 m/sec
Sight	Front—folding blade Rear—tangent with V-notch
Feed system	50 round non-disintegrating link belt
Cyclic rate	1100–1200 rpm
Manufacturer	State factories

MG42 LMG

Shortly after the start of World War II, the German Army expressed a need for a gun to replace the expensive MG34 which had proved unreliable under adverse conditions. The design of the MG42 marked a complete change in contemporary automatic weapon production. Utilizing a maximum of steel stampings the weapon had a then unique locking system.

With the bolt in the forward position, two locking rollers, situated in the front portion of the two-piece bolt, are cammed into engagement with recesses in the barrel extension by the wedge-shaped firing-pin carrier contained in the rear half of the bolt. On firing, the rollers hold the bolt in place until the barrel has recoiled a distance of 8 mm, whence inertia takes over to unlock and allow the normal reloading cycle to take place. A recoil booster muzzle attachment, using gas trapped at the muzzle to push on the barrel face assisting it rearwards, also doubles as a flash hider. Due to the high cyclic rate (1200 rpm), the barrel was made quickly detachable by placing the rear end in a yoke which, when the release lever was pushed forward, would swing the barrel out to the right of the gun and allow it to be withdrawn. A stud on the top of the rear bolt section operates a curved feed arm, pivoting at the rear of the feed cover, to shuttle the feed across and back. By the end of World War II, nearly one million had been produced since its introduction in 1942.

Some five years after the end of World War II, when the newly formed Bundeswehr was being equipped, production of the MG42 began again. Altered to take the 7·62 mm (NATO) round, and redesignated MG1, this model differed by having only very minor modifications to the bolt and feed mechanism. Some original MG42s were also converted to 7·62 mm (NATO) and designated MG2. The final version is the MG3, the main difference being that the feed has been adapted to accept not only the standard DM1 non-disintegrating link belt, but also the DM6 and US M13 non-disintegrating link belts. The ejection port has been enlarged and the barrels have chrome-plated bores.

When used with the bipod, a drum containing a 50 round belt, or box with 100 rounds, may be mounted on the side of the receiver. A tripod is available for use in the sustained fire role. A folding sight (for use against aircraft) is fitted as standard in front of the rear sight. Produced under licence in Italy for the Italian Army, the gun has been modified by the introduction of a heavier bolt and buffer spring which reduce the rate of fire down to approximately 700–900 rpm. Yugoslavia still produces the gun in its original calibre of 7·92 mm, and calls it the SARAC.

In service with the Yugoslav forces.

M65B light machine gun

Yugoslavia

Calibre	7·62 × 39 mm
Dimensions	Length 108 cm Barrel length 45·7 cm
Weight	5·2 kg
Effective range	400 m
Operations	Gas
Type of fire	Selective
Muzzle velocity	730 m/sec
Feed system	30 or 40 round box magazine
Manufacturer	State factories

Based on the standard Soviet AK47, the M65B LMG has a quick change barrel and bipod legs. It does not have the range of the RPK due to its shorter barrel and compares with the AK47 in operational characteristics.

M65B LMG

Close-up of the M65B LMG. Note the plastic pistol grip

30J light machine gun

Yugoslavia

Manufacture State factories

See the ZB26/30.

In service with some reserve forces in Yugoslavia.

Missiles and Rockets

General

The infantry in a modern war needs weapons for three main purposes: to destroy tanks, to destroy other infantry and to destroy aircraft. It makes little difference if the weapons are employed in an offensive or a defensive role so far as their design is concerned. Probably the most important of these three purposes is the destruction of enemy tanks; thus it will be found that in this review of missiles and rockets that the anti-tank weapon predominates over the anti-aircraft weapon.

The original weapon for use against a tank was the gun, which was large and unwieldy, not particularly accurate and not an infantry weapon. It soon became evident that what was required was a lighter, more mobile weapon which could be used by soldiers in forward areas of the battlefield, so the rocket was developed, and this was followed by the guided missile.

A rocket is launched from a variety of containers, often mounted on a simple support on the ground, or from the shoulder. Rockets are propelled by the jet stream from a motor or from the combustion of cordite or other explosive. The NATO standard (sic!) for rockets calls for a flight time of not more than 1·25 sec when used in the infantry role. This limits their range to under 400 m. Other armies allow a longer flight time but, in general, rockets have ranges of less than a quarter of those of guided missiles. Most rockets have hollow-charge warheads and are as capable of penetrating tank armour as are missiles. They are smaller and more portable than missiles and considerably less expensive. They are undoubtedly useful weapons, but are easily outranged by tank guns; thus they tend to be used more as a last ditch weapon fired from behind cover and, unlike a gun, make no noise or flash on being launched.

With the modern highly mobile tank it is not easy to hit the target with a rocket which, once it is launched, is committed to one trajectory only. Further, there is the sensible requirement for the infantry to outrange the enemy's tank fire. This led to the development of the missile which, because it had a much longer range, had to be guided in flight to take account of the target's movements.

Wire-guidance

The majority of anti-tank missiles in use today are wire-guided. That is to say, they unreel a length of very fine wire (or sometimes two wires) behind them and the operator uses this wire to pass electrical signals to the missile in order to steer it. The operator must have the target in sight all the time and the range of the missile is limited, both by the amount of wire it can carry and the endurance of its propulsion motor.

Wire-guidance is used because it is not subject to enemy electronic countermeasures, and is less expensive than fitting the missile and launcher with sophisticated radio or other electronic equipment. With wire-guidance, the operator must have a clear view of the target and he must remain in position for the time of flight of the missile. Further, in most systems, the operator is actually steering the missile from the moment of launch and, thus, he becomes an obvious target for the enemy. The United Kingdom, with her Swingfire system, has managed to separate the launcher and the operator by up to 100 m, and it is possible for the launcher to be behind cover, provided that the operator is in a position from which he can see the target.

The original guidance was purely optical; the operator observed the target, launched his missile, which he was able to see, and steered it manually by means of a joystick. Recently a more sophisticated method of guidance has been introduced. The missile carries infra-red flares in its tail, and at the launcher there is an infra-red sensor which measures the angle between the missile's path and the direct line of sight. If the missile deviates from this line, correcting signals are automatically passed down the wire to bring it back. All the operator has to do is to keep the target on the cross-wires of his sight. Examples of this method are the Franco-German Milan and Hot and the United States Tow.

One-man missiles

There are a number of 'one-man' missiles on the market and most of them have a range of about 200 m, which is probably sufficient as a lone infantryman is unlikely to be able, or be permitted, to pick off a tank at a longer range. The missiles can be launched either directly off a small launcher on the ground, or from the shoulder.

In the 1973 Middle East war, the Egyptians did considerable damage to the Israeli tanks with the Sagger missile, a very small anti-tank missile contained in a box no larger than a suitcase. Small weapons of this type and rockets are particularly suited to guerilla-type operations. They can be easily concealed and can do considerable damage. With the NATO nations faced with a 3–1 superiority of Warsaw Pact tanks, anti-tank weapons are assuming a supreme importance and there is an encouraging school of thought that NATO should be concentrating more on the very small missile or rocket for use by tank-hunting groups and in harassing operations.

Helicopter-mounted weapons

With NATO's forces being so much smaller numerically than those of the Warsaw Pact and with the latter's free choice of the attack point, it is difficult for infantry to be sure of being in the right place at the right time. Mobility becomes all important. Thus NATO is at last concentrating on arming helicopters with anti-tank weapons.

Although the French, and later the British, had mounted the SS11 successfully in their helicopters for a long time, it was the United States, with their Vietnam experience, who eventually took the lead in this form of warfare, and fitted a number of their Huey Cobra helicopters with the Tow anti-tank missile. These aircraft can carry two or four missiles on each side in launcher pods. The Franco-German Hot has been test-fired from an Alouette, but will probably eventually be fitted in a German B0105 or a French SA 360, either of which could carry four to six missiles, plus a crew armed with the hand-portable Milan.

The British were experimenting with the Swingfire mounted in a Puma helicopter and renamed the missile in this role Hawkswing. However the project has now been dropped. A helicopter needs a missile with as long a range as possible to enable the target to be engaged outside the range of hostile anti-tank or anti-aircraft fire.

The principle of operating helicopters in the anti-tank role is to form small battle groups of perhaps one or two reconnaissance helicopters with up to six attack machines. The group would endeavour to keep out of sight in dead ground and would send up one reconnaissance vehicle to find the enemy and to call up the attack helicopters by radio. They then literally 'pop up', deliver their missiles, and 'pop down' behind cover again as soon as possible.

With wire-guided missiles, such as Tow and Swingfire, the helicopter has to remain in position during the flight of the missile so that it can be guided to hit. Pilots would much rather launch their missiles and get behind cover with the least delay; hence the requirement for a self-homing missile which would not require pilot guidance and a number of such systems are under development.

Vulnerability of helicopters

Helicopters are, of course, particularly vulnerable to small anti-aircraft missiles at the low heights at which they must operate, and in particular to the one-man shoulder-launched missile. Such missiles did considerable damage to Israeli aircraft in the Yom Kippur war of 1973. Nearly all missiles of this type home on the heat emitted from the aircraft's engines, and all that the aimer has to do is to point the missile in the general direction of the aircraft, press a button and the missile does the rest. An exception to this is the British Blowpipe, which is automatically guided for the first two or three seconds of its flight and then manually guided by the operator, who has to continue to track the target through his sight. As a result Blowpipe is of limited use at night.

One man missiles do not have a great range—something of the order of 3 km is normal—but they are invaluable to the infantryman undergoing direct aircraft attack for dealing with helicopters and for clandestine work in the vicinity of enemy airfields. Being small, they might not necessarily shoot down an aircraft, but they should certainly deliver a powerful enough explosion to cause damage. One of the problems with hand-held missiles is the difficulty of carrying a supply of missiles, and some missile and rocket launchers are now designed to fire only one or two missiles/rockets and are then abandoned. It follows that to produce an effective volume of fire it is necessary to arm a large number of soldiers with a large number of launchers, and this raises other problems.

Standardisation

As will be seen, the NATO countries and France have developed some 10 vehicle or ground-launched anti-tank missiles and eight hand-held ones, plus numerous different types of rockets. Many of the missiles perform almost identical functions and it seems a great waste of time and effort to have done so much work, not to mention the production which is now being carried out, when one, or at the most two, standard weapons could have sufficed. The Franco–German Hot, the United States Tow and the British Swingfire all have very similar performances, and the German Cobra, the United States Dragon, the Franco–German Milan, the Italian Mosquito are all weapons designed for the same purpose—one-man launching.

Standardisation on such issues inside NATO has never really got off the ground; anti-tank and ground-to-air missiles are both examples of the waste of effort and money caused by the inability of NATO nations to agree amongst themselves. It is a different story amongst the nations of the Warsaw Pact.

Sophisticated infantry weapons are likely to increase in numbers and complexity and, if mobility is to be NATO's keyword, the infantryman may find himself in Turkey at one moment and in Norway at the next. He would like to find stocks of weapons in which he has been trained in both countries. At present this is highly unlikely, unless the NATO countries drastically revise their procurement methods.

T21 (Tarasnice) anti-tank gun

Czechoslovakia

Calibre	82 mm
Length	147 cm
Weight	20 kg (with carriage)
Range	300 m (direct) 2800 m (indirect)
Sight	Optical and battle
Ammunition	HEAT
Effect	Penetrates 230 mm of armour plate
Manufacturer	State factories

The Czech T21 recoilless anti-tank gun is a very versatile weapon, capable of being fired from its small two-wheel carriage, from the shoulder or from a fixed mount on the Czech tracked OT62 armoured personnel carrier, and in both direct and indirect roles.

It is a smooth-bore weapon, firing a shaped-charge HEAT fin-stabilised round, which is breech-loaded, and fired electrically—as with the P27 anti-tank grenade launcher, the trigger action works a magneto. To cope with the direct and indirect fire capability, the T21 is equipped with ordinary iron sights graduated in 50 m intervals and also a telescopic sight graduated at 100 m intervals.

The HEAT round, 2·1 kg in weight, can penetrate up to 230 mm of armour at an effective range of 300 m if the target is moving, and up to 650 m if the target is stationary. Rate of fire is about 5 rpm, but this is adversely affected by the absence of a used round extractor in the breech, which has to be cleared manually with the aid of an asbestos glove.

The T21 is usually transported in a truck or armoured personnel carrier, but can be towed for short distances using the long tubular towing handle, normally kept folded over the top of the tube.

In service with the Czechoslovak, some Eastern and the Vietnamese forces.

T21 (Tarasnice) 82 mm RCL

M59 and M59A anti-tank gun

Czechoslovakia

Calibre	82 mm
Length	381 cm
Weight	386 kg
Range	7550 m
Sight	Optical (PB04K)
Ammunition	HEAT and HE
Effect	Penetrates 250 mm of armour plate
Crew	5
Manufacturer	State factories

The Czech M59A recoilless gun is unique in that it is the only weapon of this type used by Warsaw Pact countries to have a spotting rifle, the semi-automatic 12·7 mm ZH59. Although designed basically as an anti-tank gun for use in the infantry regiment, the very long barrel and provision of both HEAT and HE rounds enable this weapon to be used as a field gun. The M59A is distinguished by its long barrel, spotting rifle and the small flared blast-shield behind the breech and is visibly different from the original M59 only by reason of the ribbing on the breech.

The PB04K sight, mounted on the left-hand side in front of the breech, is employed for both the indirect and

M59A

direct fire roles. The 6 kg projectile is breech-loaded and the HEAT round can pierce up to 250 mm of armour. Both versions of this gun are usually mounted on a light two-wheeled, rubber-tyred carriage, towed by a vehicle, and can also be moved easily for short distances manually by the small hand-towing bar on the muzzle. The weapon has also been found mounted on armoured personnel carriers and light trucks, but in all cases the rate of fire is around 6 rpm, and a crew of five are needed to serve the gun.

In service with the Czechoslovak and some Third World countries' forces.

P27 (Pancerovka) anti-tank free flight missile launcher

Czechoslovakia

Calibre	45 mm (missile 120 mm)
Length	103 cm
Weight	6·4 kg
Range	100 m
Sight	Tangent leaf and hooded post
Ammunition	HEAT
Effect	Penetrates 250 mm of armour plate
Manufacturer	State factories

The Czech P27 launcher was produced at about the same time as the Soviet RPG2 which it resembles. The

P27 (Pancerovka) 45 mm missile launcher

P27 incorporates a tube, 103 cm long (sometimes fitted with a small, flared blast-shield at the rear end) and an insulated guard covering most of the rear half, a pistol-grip with trigger and electrical firing mechanism, large fore and back sights and a bipod.

The bipod and the large size and sophistication of the sights are the main differences between this weapon and the RPG2. The fin-stabilised, shaped-charge HEAT round is loaded by hand into the muzzle and is electrically fired—the trigger action powers a magneto. The missile will penetrate up to 250 mm of armour, but can only be used effectively against a moving vehicle at about 80 m. Beyond 100 m targets must be limited to men or soft-skinned vehicles. Up to five rounds can be fired in one minute. The front sight is a hooded post, and the rear sight a tangent leaf graduated at 50 m, 75 m, 100 m, 125 m and 150 m, the two folding together for carriage, for which a sling is provided.

Close-up of P27

In service with the Czechoslovak, some Eastern and the Vietnamese forces.

B10 anti-tank gun

USSR

Calibre	82 mm
Length	170 cm
Weight	87·5 kg
Range	4400 m (effective 400 m)
Sight	Optical (PB02)
Ammunition	HEAT and HE
Effect	Penetrates 250 mm of armour plate
Crew	4
Manufacturer	State factories

The Soviet-made B10 (or RG82) recoil-less gun replaced the SPG82 rocket launcher as the battalion anti-tank weapon in about 1950 and, although now largely replaced in its turn by the SPG9, is still to be found in some satellite countries and China. Considerably heavier than its predecessor, the B10 is equipped with the more modern PB02 sight for both direct and indirect fire and using a 7·3 kg, shaped-charge, fin-stabilised HEAT round, and an 8·2 kg HE round, is capable of much more effective fire—the HEAT round can penetrate up to 300 mm of armour at 400 m.

The B10 is breech-loaded and percussion-fired, the trigger and bolt being accommodated in a pistol-grip located above and to the right of the tripod mount, whilst the optical sight is on the opposite side in front of a rectangular, perforated guard. The gun is usually transported in a truck or armoured personnel carrier,

Soldiers pulling a B10

82 mm B10 (RG82)

but can be towed by its four-man crew on its small two-wheeled carriage using the conspicuous towbar fitted below the muzzle. The small castor wheel below the towbar is designed to prevent the muzzle touching the ground while the gun is being manoeuvred or set up for firing on the tripod. The latter can be adjusted to give either a low silhouette or a better field of view, but the weapon can, if necessary, also be fired while still on the wheeled mount. A rate of fire of 6–7 rpm is attributed to the B10.

In service with the Angolan, Bulgarian, Chinese, East German, some Near Eastern and the Vietnamese forces.

B11 (or RG107) anti-tank gun

USSR

Calibre	107 mm
Length	350 cm
Weight	305 kg
Range	6650 m (450 m effective)
Sight	Optical (PB04)
Ammunition	HEAT and HE
Effect	Penetrates 360 mm of armour plate
Crew	5
Manufacturer	State factories

The B11 (or RG107) is the biggest recoilless gun manufactured by the USSR, having been developed (along with the B10) from knowledge gained from captured United States weapons during the Korean War. Whilst the B11 is designed mainly as an anti-tank gun, able to pierce 350 mm of armour at 450 m, it is light and mobile enough on the two-wheeled carriage to be manoeuvred in battle by its five-man crew using the light towbar attached below the muzzle. Although the B11 can, like the smaller but similar B10, be fired from its wheeled mount, the motor-car-type wheels are awkward and are often removed while the gun is raised on its tripod.

The two main recognition features of this weapon are the tripod which, while travelling, is spread with the front leg attached to the tube and the two rear legs raised and carried at an angle to the rear, and the limber loop above the muzzle by which the gun is connected to its towing truck.

The B11 is breech-loaded, using a 16·8 kg shaped-charge, fin-stabilised HEAT round or a 21·3 kg HE round, both types of ammunition being served by the same PBO4 direct and indirect fire sight, which is mounted high and centrally on the gimbal mounting of the barrel. The USSR no longer uses this, or any similar calibre recoilless gun, in its first-line divisions.

In service with some Near Eastern, the Vietnamese and Warsaw Pact forces.

107 mm B11 (RG107)

SPG9 anti-tank gun

USSR

Calibre	73 mm
Length	200 cm
Weight	60 kg
Range	1200 m
Sight	Optical
Velocity	700 m/sec
Ammunition	HEAT, rocket assisted
Effect	Penetrates 390 mm of armour plate
Crew	3
Manufacturer	State factories

The SPG9 recoilless gun is a breech-loading weapon which is now gradually replacing the B10 in several Warsaw Pact countries. The SPG9 has a long, clean tube with a carrying handle (to allow manoeuvring while the tube is hot) set about 25 cm from the muzzle. It is mounted permanently on a tripod mount, of which the legs fold forward for packing. No wheels are provided, as the weapon is usually transported in an armoured personnel carrier, although it is light enough to be carried by two men. The missile, probably a shaped-charge HEAT round, is similar to that used by the RPG7V, having an initial propellant cartridge and fin-stabilisation. An optical sight is located on the left of the tube above the mount, and the foresight is also on the left attached above the forward carrying handle.

In service with the Polish, Soviet and most other Warsaw Pact forces.

75 mm SPG9

RPG2 anti-tank free flight missile launcher

USSR and China

Calibre	40 mm (launcher) 82 mm (missile)
Length	145 cm
Weight	2·7 kg
Range	150 m
Ammunition	HEAT
Effect	Penetrates 180 mm of armour plate
Crew	1
Manufacturer	State factories

The original Soviet RPG2, now obsolescent, is a very simple and cheaply produced, infantry section anti-tank grenade launcher. It is a later version of the RPG1, developed from the German World War II Panzerfaust system. It consists of a cylindrical tube with an awkward-looking pistol-grip containing the trigger, bolt and firing pin, a small foresight and tall rear sight, and a wooden covering round most of the tube behind the pistol-grip. The reason for the large rear sight, which is graduated for 50 m, 100 m and 150 m, is the high trajectory of the 2·7 kg shaped-charge HEAT round, which has a maximum anti-tank range of only 100 m, and a penetration capability of 180 mm of armour. The fin-stabilised round is muzzle-loaded and percussion-fired. A safety-catch is provided. The weapon can be fired stand-

40 mm RPG2

ing, kneeling or lying but only from the right shoulder, as there is a gas-ejection port on the right side of the tube above the pistol-grip. Later models have a small, flared blast-shield on the open rear end of the tube and can be fitted with the Soviet NSP2 infra-red night sight. Rate of fire is about 4 rpm.

The RPG2 is similar to the Czech P27, but has simpler sights and no bipod. The Chinese communists make a copy of the RPG2 which they call the anti-tank grenade launcher Type 56, and in North Vietnam the weapon is known as the B40.

In service with the Angolan, Chinese, Vietnamese and some Warsaw Pact forces.

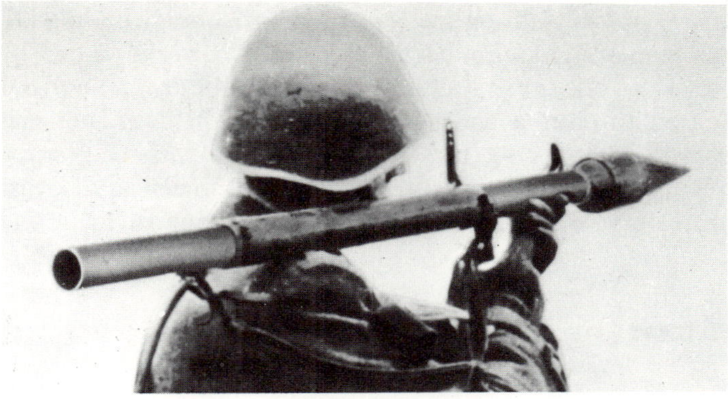

RPG2 with missile

B40 anti-tank free flight missile launcher

USSR and China

Manufacturer	State factories

See the RPG2.

In service with the Vietnamese forces.

RPG7V, RPG7, RPG7D anti-tank free flight missile launcher

USSR

Calibre	40 mm (launcher) 85 mm (missile)
Length	100 cm
Weight	8·5 kg (loaded)
Range	300–500 m
Velocity	300 m/sec
Sight	Telescopic ($\times 2\frac{1}{2}$) and battle
Ammunition	HEAT
Effect	Penetrates 320 mm of armour plate
Manufacturer	State factories

40 mm RPG7

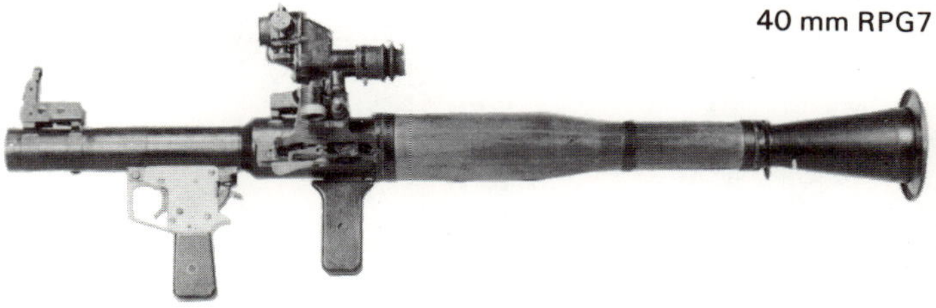

40 mm RPG7V

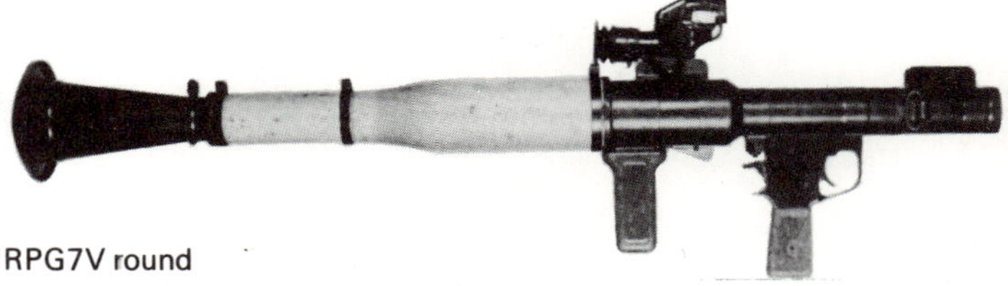

RPG7V round

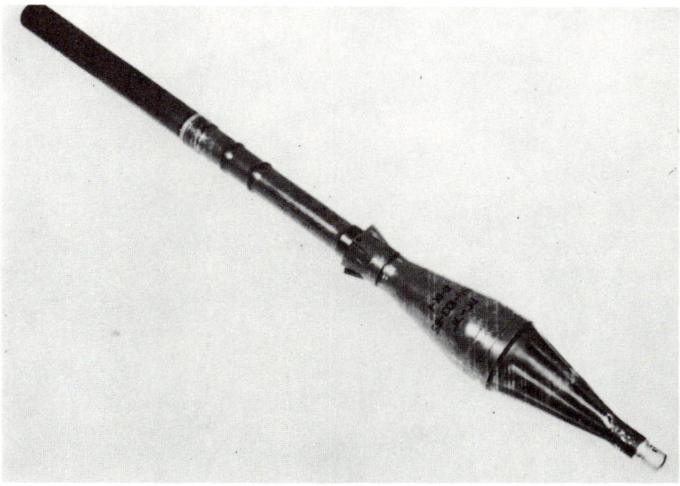

The Soviet RPG7V anti-tank grenade launcher is the replacement for the RPG2, affording improved range and penetration. Although still operating on the basic German Panzerfaust principle of a tube open at both ends with a pistol-grip and trigger-firing mechanism, there are several technological improvements over RPG2, some of which convey a marked difference in the new weapon's appearance. These include the blast-shield of the RPG7V which is large and conical; to provide greater thrust the centre tube section covered by part of the wooden heat shield has been expanded; to improve handling there are now two pistol-grips; there are now two relatively small iron sights graduated every 100 m from 100 to 500, and in addition a $\times 2\frac{1}{2}$ telescopic sight can be used as well as the NSO2 infra-red night sight.

The 1·7 kg rocket-assisted, fin-stabilised, shaped-charge HEAT round is muzzle loaded and percussion fired, although the projectile itself utilises a piezo-electric fuse which self-destructs at 900 m and the PRO7 telescopic sight can be illuminated for night use. The RPG7V missile is launched from the tube by the normal propellant but a short distance from the launcher the rocket motor cuts in, thus giving the missile greater accuracy from the resultant increased thrust and lower trajectory; 320 mm of armour can be penetrated at the much increased range. The original PRG7 was slightly larger than the RPG7V, and fired a slightly smaller projectile. RPG7D is the paratroop version of RPG7V, and can be divided in two for compact transportation.

In service with the Egyptian, Iranian, Syrian, Vietnamese and Warsaw Pact forces, and the IRA.

Swatter (AT2) anti-tank guided missile

USSR

Dimensions of missile	Length 113 cm Diameter 13 cm
Weight	Missile 26·5 kg
Range	600–2500 m
Sight/aiming	Optical
Operation	Radio command to line of sight
Ammunition	HEAT
Manufacturer	State factories

Normally mounted in a vehicle, Swatter is one of the few anti-tank missiles in the world which is not wire-guided. The operator tracks the missile through a sight and is assisted by two flares in its tail. He then has to guide the missile to the target by issuing radio commands by means of a joystick, a system which the Western nations discarded years ago.

The command radio link has two channels in case one is jammed. Normally four missiles are mounted in one vehicle.

In service with the Egyptian, Syrian and Warsaw Pact forces.

Sagger AT-3 anti-tank guided missile launcher

USSR

Dimensions of missile	Length 87 cm Diameter 12 cm
Weight	11 kg
Range	500–3000 m
Sight/aiming	Optical, with joystick
Operation	Automatic gathering, command by wire
Ammunition	HEAT
Actuation	Impact
Effect	Penetrates 400 mm of armour plate
Crew	3
Manufacturer	State factories

Sagger

The infantry version of the Sagger anti-tank guided missile is operated by a team of two men, normally from an unprepared site on the ground, but the weapon can be fired from an armoured personnel carrier or truck. The weapon kit consists of a two-part missile (HEAT

Sagger in action. (Missile is behind the optical sight)

warhead and body), carried in a glass-fibre case which serves as the launch pad, and a control box. As the missile is of lightweight construction, the crew can carry several at once. For normal operation the missiles, which are wire-guided, are set up, facing the direction of enemy advance, on the light launch rail contained in the carrying case. Maximum range is about 3000 m and the HEAT warhead will penetrate up to 400 mm of armour at 0° impact angle on impact at any distance up to that range. The control box incorporates activation and flight control switches set in front of the optical sight. On being fired by electrical impulse from the activation switch, the missile is gathered in the optical sight and guided on to the target by manipulation of the flight control switch. Although this system is very reliable, the operator needs constant practice to remain efficient.

A similar version of this missile is also used by the countries listed, in fixed mounts on various armoured vehicles.

In service with the Angolan, Egyptian, Syrian, Warsaw Pact and Yugoslav forces.

A Warsaw Pact soldier setting up a Sagger

RB57 (M57) anti-tank free flight missile launcher

Yugoslavia

Calibre	44 mm (missile 90 mm)
Length	96 cm
Weight	8·3 kg
Range	1200 m (effective 400 m)
Sight	Folding tangent leaf and battle
Ammunition	HEAT
Effect	Penetrates 270 mm of armour plate
Manufacturer	State factories

The Yugoslav RB57 anti-tank grenade launcher was produced in 1957 as a result of research into the German Panzerfaust system and has subsequently replaced the United States World War II launchers used until then by the Yugoslav forces. The RB57 has thus a certain resemblance to the Soviet RPG2, but is much more sophisticated and is almost comparable in performance to the RPG7V (but having a longer maximum range).

The RB57 is, however, considerably heavier and the

1·8 kg shaped-charge, fin-stabilised HEAT round is not rocket assisted, having a maximum effective range against a stationary target of only 400 m, which is reduced to 200 m if the target is in motion, and only 270 mm of armour can be penetrated. Rate of fire is 4 rpm. The projectile is muzzle-loaded and fired by means of a primer set (at a right angle to the direction of flight) in the lower side of the tail boom. The primer is detonated by a spring-loaded striker housed in the cylinder set in front of the pistol-grip assembly. The striker is activated and automatically reset by a double-action trigger. The standard sights are of the folding tangent leaf type, graduated every 100 m up to 400 m and, for fast close-range action, a small battle sight graduated to 200 m is provided. The weapon has occasionally been seen with an optical sight fitted.

In service with the Yugoslav forces.

44 mm RB57

M60 anti-tank gun

Yugoslavia

Calibre	82 mm
Length	220 cm
Weight	122 kg
Range	4500 m (effective 500 m)
Sight	Optical (PTDM60) and battle
Ammunition	HEAT and HE
Effect	Penetrates 220 mm of armour plate
Crew	5
Manufacturer	State factories

The Yugoslav M60 recoilless gun is similar to the Soviet B10 of the same calibre, but is longer and heavier and has a longer maximum anti-tank range. The M60, like the RB57 anti-tank grenade launcher, is one of three anti-tank weapons (the third is the M65 105 mm recoilless gun) developed by the Yugoslavs to take the place of the United States anti-tank weapons on which they had relied since World War II.

A breech-loading smooth-bore weapon, the M60 uses a 4·3 kg shaped-charge, fin-stabilised HEAT round which will penetrate 220 mm of armour at ranges up to 500 m, and also a HE projectile up to the maximum range of 4500 m. This round resembles an artillery shell at first sight, but the finned boom is protected by a perforated sheet casing which is released on launching.

82 mm M60

The optical PTDM60 sight is set level with the tube on the left side and is graduated up to 1500 m, but additionally, as in the case of the RB57, there is a battle sight fitted for quick action.

Normally the M60 is towed behind a light truck on its light, curved, tubular, two-wheeled carriage, but it can also be either towed easily by its five-man crew by the double-handed towbar below the muzzle, or quickly broken down for pack transport by the crew or by mule. This is a common feature of most Yugoslav-made weapons.

In service with the Yugoslav forces.

M65 anti-tank gun

Yugoslavia

Calibre	105 mm
Length	415 cm
Weight	280 kg
Range	6000 m (effective 600 m)
Sight	Optical
Ammunition	HEAT
Effect	Penetrates 330 mm of armour plate
Manufacturer	State factories

105 mm recoilless M65

Whereas the USSR has now given up large-calibre recoilless guns, Yugoslavia, with its problem of finding effective weapons of sufficient manoeuvrability to be of use in mountainous terrain, has in addition to the two lighter anti-tank weapons, the RB57 and the M60, developed its own large recoilless rifle, the M65. Among communist countries' weapons, the M65 is unique in that it is fitted with a 12·7 mm spotting machine gun, the UB, used at ranges up to 600 m (the M65's maximum effective anti-tank range) and employing a 20 round belt, fired either automatic or single shot. The M65 is a breech-loading weapon with a very long rifled barrel, allowing the gun to fire its HEAT projectiles a distance of 6000 m. HE rounds have not yet been seen with this weapon. Six rounds can be fired in one minute and the shaped-charge warhead will pierce 330 mm of armour plate.

The gun is mounted on a tripod, the two wheels attaching to the base of two of the legs, whilst the third leg extends to the rear and also acts as a towbar. A small, square shield which protects the firer's face from blast is fitted in front of the optical sight. This weapon can travel in the normal manner towed by a truck or armoured personnel carrier, or can be broken down for carriage by mule.

In service with the Yugoslav forces.

SA-7 anti-aircraft guided missile

USSR

Length of launcher	135 cm
Weight	10·6 kg
Range	Out to 3700 m (possibly 4500 m)
Speed	Mach 1·5
Operation	Heat-seeking/infra red
Sight	Battle
Ammunition	HE
Actuation	Impact
Crew	1
Environment	Pan climatic
Manufacturer	State factories

The SA-7 heat-seeking, low-level surface-to-air missile was developed by the USSR to counter the threat of ground-attack aircraft and helicopters armed with anti-armour and anti-infantry weapon systems. It has proved its value against such targets in Vietnam and during the

1973 Arab–Israeli war but, because of its lack of radar target acquisition facilities, the hand-held system of operation and the small size of the warhead, it is not effective against larger and faster aircraft. Additionally, the essentially heat-seeking nature of the projectile means that it must attack the exhausts of the target and can be diverted by ejection of flares and other manoeuvres, such as flying below 150 m. The system consists of a launching tube about 150 cm in length, fitted with a solid shoulder-pad behind the pistol-grip firing mechanism, a thermal battery forward of the pistol-grip and simple battle-sights mounted above the left side of the tube. A distinctive heavy strengthening band is fitted round the muzzle, and a canvas sling is provided for carrying purposes. The SA-7 system equates to the United Kingdom Blowpipe and the United States Red Eye and is alternatively referred to as Strela (Russian—'Arrow').

Because of its easy portability, this missile system is being increasingly encountered with Russian-supplied guerrilla forces in Africa and the Middle East, particularly in Eritrea, Oman and Rhodesia, whilst the Vietnamese are reliably reported to have a version capable of reaching a ceiling of 4500 m. A form of manual guidance has now been added.

SA-7 in action

In service with the Angolan, Cuban, Egyptian, Syrian, Vietnamese, Warsaw Pact and Yugoslav forces.

Mortars

The mortar is a weapon in which the main recoil force is transmitted directly to the ground through a base plate attached to the rear of the barrel. It is this lack of a recoil system which distinguishes a mortar from all other guns and howitzers which fire explosive shell. It is no new invention. Mortars have been in service since the earliest guns—the first recorded instance of their use being in the Siege of Constantinople in 1451, when one or two large guns were upturned and used to throw stone balls in a high trajectory on to the decks of ships. After that date, mortars continued as an artillery, or ship-board, piece of ordnance and they were all large and heavy.

Infantry had no use for these early mortars, and indeed they had hardly any need of them since the guns were always close to them supplementing the fire of the muskets with round shot and grape. The arrival of the modern rifled guns with their long range brought a change to the intimate close support provided by the smooth-bore gunner and, in World War I, the infantry hardly saw its supporting artillery at all. It was this war which saw the re-introduction of the mortar to the battlefield, its operators being mainly infantrymen. When the opposing forces were dug in within a few hundred yards of each other, a weapon was needed which would throw projectiles, and preferably big projectiles, into the enemy trenches. For this the mortar was ideal and in no time a number of curious—and dangerous—equipments were being made to throw containers of high-explosive across no-man's-land. The high trajectory of these mortars meant that their projectiles fell in an almost vertical straight line and if the range had been correctly estimated no trench was proof against them. But they were still heavy and clumsy, and the rate of fire was pitifully slow.

Sir Wilfred Stokes, who was the managing director of a firm of manufacturers of agricultural machinery, changed all this in 1915. He produced the prototype of what has become the standard infantry mortar throughout the world. The Stokes mortar was a simple tube of good quality steel mounted on a flat steel base-plate and supported at the muzzle end by a bipod which was adjustable in elevation and azimuth by means of screws. At the breech was a fixed firing pin, while the propellant was contained in a cartridge in the tail of the bomb. To fire the mortar the bomb was dropped down the barrel tail first. By this means a very high rate of fire could be achieved since the only delay between rounds was the time needed for the bomb to fall down the barrel and for the loader to put another in at the muzzle. All but a very few mortars use the Stokes principle today.

After World War I the use of Stokes-type mortars spread rapidly and, by 1939, all the first-line armies had one or more types of mortar in service. World War II gave a great impetus to their use and manufacture since they provided a useful source of close support fire for infantry, and the control and handling of them passed into the hands of infantry battalions, except for the large versions which remained artillery pieces in all armies.

All light and medium mortars can be manpacked and so can be taken nearly everywhere a man can move. They are reasonably cheap to make and to maintain. The rate of fire can be very high for short periods, as many as 20 rounds per minute being achievable: a platoon of 81 mm mortars can put down a heavier weight of metal on the target than can a battery of field guns—though only for a short time. Concealment of a mortar is generally easy, and the control of the fire is straightforward and does not require complicated survey techniques as with a gun. By being sited inside the infantry battalion area, the response to fire requests is rapid and the high angle fire allows targets behind cover to be engaged.

The disadvantages are less apparent, but exist all the same. The range of most mortars is short for present day engagements, and efforts to increase the range generally end by making the mortar larger and heavier and more like a howitzer. Until recently, accuracy was poor because there had to be 'windage' around the bomb when it was in the barrel to allow it to drop down quickly on to the firing pin. This 'windage' also allowed the propellant gas to escape round the bomb on firing and gave rise to varying ballistics and a spread of shot on the target. However, such loose-fitting bombs were not unduly expensive to make and so the inaccuracy could be compensated to some extent by firing more of them. This situation has now partially changed. Modern mortars still use a loose-fitting bomb—but with a plastic driving band which expands on firing to fit the bore so that accuracy is now very good.

Although the mortar is reasonably cheap, the ammunition nowadays is not. Modern bombs are made to very fine tolerances and so are expensive. A typical medium mortar will cost about 100 times more than one round of HE or, to put it the other way round, 100 bombs equal the price of the mortar which fires them. With light mortars the cost of the ammunition is proportionately higher. A good-quality light mortar with an accurate sight will only cost the same amount of money as 40 or 50 of its bombs. The reason is that although the mortar is cheaper to make than the medium variety, the ammunition is not much different in total cost. Modern bombs are made to fine tolerances and so are expensive; they take a good deal of semi-skilled factory labour to produce. Moreover, they are heavy and awkward to carry and, although three men can carry a medium mortar on their backs, they may have to be followed by a long line of other men carrying the ammunition. Helicopters, however, are reducing the need for manpacking on the modern battlefield.

Mortars are generally classified in three sizes, light, medium and heavy. Light mortars appeared in the 1920s to provide intimate support to the infantry platoon and company and to provide a greater range and effectiveness than could the various rifle-mounted grenade launchers which came into service in World War I. Most of these light mortars survive, though their use is declining as other weapons replace them. The simplest light mortar was, and still is, the British 2 in. This is a piece of straight tube with a minimal base plate and a simple trigger mechanism, as the bomb is not fired by the Stokes principle. There is no sight and aiming is by eye and judgement. The range is about 500 m and the consistency good. At the other end of the scale is the 60 mm family developed from a French design. Some of these weigh almost 18 kg and have a maximum range of 1500 m. Not only are they too heavy for a platoon, but the maximum range is too great for a platoon's needs and is generally wasted. More than 1000 m range requires some sort of fire controller to correct the fall of shot, and this is absent in the 60 mm field. Most of these complicated light mortars have now fallen into disuse, though some remain in the armies of the emergent nations. The 2 in is retained in the British and Commonwealth armies for smoke and illumination purposes but is due to be replaced by an updated version in a calibre of 51 mm with an increased range out to 700 m and better accuracy. It will use an obturating ring on all its bombs and will have a proper sight together with a monopod.

Medium mortars are generally battalion weapons and their calibre varies between 70 mm and 90 mm. In general, 80–82 mm covers the majority of all medium mortar calibres around the world. The Western nations tend towards 81 mm, and the Soviet Bloc 82 mm. It has been said that this extra 1 mm gives the Soviets the advantage in that they can fire Western bombs whereas the reverse is not possible. Just how much this matters is very questionable. All these medium mortars have roughly similar characteristics and performances. The maximum range is about 4000 m, the weight of the bomb is about 4·5 kg–5·4 kg, the total weight of the mortar is about 45 kg and it breaks down into three manpack loads. In most armies there is a platoon of six to the battalion and in mechanised battalions the mortars are carried in personnel carriers with open tops so that the mortar can be fired from the vehicle without it having to be dismounted. These mechanised mortars are very formidable since their mobility is only limited by their vehicle and the crew can fire from what is virtually a prepared position within seconds of halting.

An interesting adaptation of the medium mortar has been produced by the French firm of Thompson-Brandt who now have a mortar-gun. This is nothing more than a breech-loading smooth-bore gun firing mortar bombs with a slightly higher charge out of a barrel which is longer and stronger than the mortar. The gun is meant to be mounted in an armoured vehicle and has a simple recoil mechanism. While not a true mortar, the idea is directly derived from the well-known Thompson-Brandt range and now some of these higher velocity bombs are being tried in the infantry mortars that the firm sells, in order to improve the performance, which they do. Some of them confound all precepts of mortaring by flying supersonically in the early stages of the trajectory and to stabilize themselves they have pop-out fins which project into the airstream, at the same time giving a slow rate of roll.

There is still plenty of scope for research in mortar bombs. All bombs are a fine compromise between stability, effectiveness, range and finally cost. To make an accurate bomb in 81 mm calibre which will fly true and yet give a useful effect at the target is extremely difficult. To design it requires a team of experienced ballisticians supported by pyrotechnic chemists, propellant designers and quite expert production managers in the actual factories. The costs are frightening, and the final price of a bomb is quite close to that for a shell; indeed both are made by roughly the same processes. Any designer who can reduce the cost by 10%, or even less, is making a substantial contribution to the budget. If the effectiveness can be enhanced at the same time, then this is a major step forward. Finally, if the range can be improved without altering any of the other characteristics, then a minor miracle has happened. Some of the best brains in the armament world are looking for this minor miracle at the moment; few are near to finding it.

The modern medium mortar probably offers the best compromise between weight, bulk and effectiveness for all of the mortar family and it is no accident that there is a great similarity in their sizes. Heavy mortars, on the other hand, are less popular and except for special purposes they are not so attractive as the medium. A heavy mortar is generally taken to be one that is over 100 mm in calibre. The most usual calibre is 120 mm and there are several different makes of this size.

The United States still has a World War II 4·2 in (about 107 mm) rifled mortar in service, though it seems likely that it will be phased out soon. There are several different makes of 120 mm mortars in service, the most prolific manufacturers being Tampella in Finland and Thompson-Brandt in France; but these heavy mortars are too big and clumsy for the infantry, except in rare circumstances, and they are given to the artillery who usually use them as light guns. Most armies which employ them man them with heavy mortar companies and deploy them as short range artillery in support of infantry, or use them for special purposes such as laying heavy concentrations of chemicals or smoke.

M48 medium mortar

Czechoslovakia

Calibre	82 mm
Barrel length	135 cm
Baseplate dimensions	63 cm diameter
Weight in firing position	60 kg
Carriage weight	60 kg
Range	4000 m
Rate of fire	8–10 rpm
Operation	Automatic
Sight	Optical
Ammunition	HE (3·3 kg)
Crew	3
Manufacturer	State factories

The Czech 82 mm mortar (M48) is another weapon developed with guerrilla warface in mind. It breaks down easily into three man-pack loads and incorporates the novel feature of a tiny fixed baseplate which allows the normal circular baseplate to be dispensed with on hard ground. The carrying handle on the muzzle cover aids transportation, especially when the mortar has to be moved, hot, in action. The elevating and traversing mechanism and sights employed, for ease of manipulation, are larger than those of most other mortars of this size.

In service with the Czechoslovak and various South East Asian and Middle East forces.

MO-81-61L medium mortar

France

Calibre	81 mm
Barrel length	145 cm
Barrel weight	14·5 kg
Bipod weight	12·2 kg
Baseplate weight	14·8 kg
Weight in firing position	41·5 kg
Range	5000 m
Rate of fire	Variable
Operation	Automatic
Sight	Cross-level collimator
Ammunition	Mk 61 (with V19P fuse)
	Mk 61 coloured HE
	Mk 61 smoke
	Mk 68 illuminating (3·5 kg)
	Mk 61 practice
Crew	3
Manufacturer	Hotchkiss-Brandt

Hotchkiss-Brandt 81 mm mortar (long barrel)

This 81 mm mortar has two barrel lengths, the short designated MO-81-61C and the long MO-81-61L. The weight of both versions is sufficiently low for them to be man-packed in three loads, or carried as a single load by a mule. The short-barrelled version fires the M57D range of ammunition (3·3 kg), consisting of HE, coloured explosive smoke and practice rounds, and the Mk 62 illuminating round (3·1 kg).

The Mk 61 range of ammunition (4·3 kg) is designed for the long-barrel mortar and consists of bombs similar to those of the M57D range. However, both mortars can fire both types of ammunition, taking range table conditions into account.

In service with the Argentinian, French (MO-81-61C), Swedish and Yugoslav forces.

M37 (new type) medium mortar

USSR

Calibre	82 mm
Barrel length	122 cm
Baseplate dimensions	50 cm diameter
Barrel weight	19·6 kg
Bipod weight	20·1 kg
Baseplate weight	21·3 kg

Weight in firing position	56 kg
Range	3040 m (minimum 100 m)
Rate of fire	15–25 rpm
Operation	Automatic
Sight	MPM-44
Ammunition	HE (3·05 kg) and smoke
Crew	5
Manufacturer	State factories

M37 82 mm mortar

M37 82 mm mortar being positioned for firing

Of the several 82 mm mortars produced by the USSR, immediately before and during World War II (the M37, M41, M42/43 and M43), only the M37 has survived, modernised, to be the standard Warsaw Pact 82 mm mortar. There is little variation in the models of this mortar manufactured by the different communist countries, although some, but not the USSR, do produce an ingenious two-wheeled carriage which also transports the ammunition. The Chinese copy of the weapon is the 82 mm mortar type 53.

Whilst the 82 mm mortar is now rarely encountered in infantry units of the Soviet Army, being found mainly in airborne units, this mortar is still widely used by the infantry of the other Warsaw Pact countries, communist China and Albania. In the Soviet Army the infantry mortar is now the 120 mm mortar M43. The 82 mm mortar M37 (new type) and, indeed, all of its predecessors can fire the 81 mm mortar ammunition manufactured in many western countries.

In service with the Soviet, all other Warsaw Pact, Afghan, Albanian, Chinese, Indonesian, Iraqi, Syrian, Vietnamese and Yugoslav forces.

M107/M38 heavy mortars

USSR

Calibre	107 mm
Barrel length	167 cm
Bipod length	120 cm
Baseplate dimensions	94 cm diameter
Weight in firing position	170 kg
Carriage weight	340 kg
Range	6300 m (minimum 800 m)
Rate of fire	10–15 rpm
Operation	Manual or automatic
Sight	Optical
Ammunition	Heavy HE (9 kg), light HE, smoke and chemical (7·9 kg)
Crew	6
Manufacturer	State factories

The 107 mm mortar of both marks, M38 and M107, is a scaled-down version of the Soviet 120 mm mortar, reduced in weight and size to suit its normal role, as the regimental mortar of mountain units. The mortar is usually transported by pack animals in difficult terrain, but is towed on a two-wheeled carriage over normal roads. For animal-packing, the weapon breaks down into three loads, barrel, bipod and baseplate.

The M107 is the improved model of this mortar and has replaced the M38 in the Soviet Army, although the earlier version is still to be found in the non-Soviet Warsaw Pact countries and in South-East Asia.

In service with the Soviet, some other Warsaw Pact and Vietnamese forces.

M38 heavy mortar

USSR

Calibre	120 mm
Barrel length	185 cm
Baseplate dimensions	100 cm diameter
Weight in firing position	170 kg
Carriage weight	340 kg
Range	5700 m (minimum 400 m)
Rate of fire	12–15 rpm
Operation	Manual or automatic
Sight	MP-41/MP-42 collimating sight
Ammunition	HE (15·4 kg), smoke (16 kg), incendiary (16·7 kg)
Crew	5 or 6
Manufacturer	State factories

The 120 mm mortar M38 was the standard Soviet infantry mortar until it was replaced recently by the 120 mm M43. A unique design when originated, the mortar can be quickly broken down into three parts, barrel, bipod and baseplate, for movement over short distances. For normal travel the whole weapon folds together and is towed on a two-wheeled carriage or, if necessary, can be animal-packed in its three parts. The mortar can be either drop- or trigger-fired by lanyard, as required, and an anti-double-loading device is provided for attachment to the muzzle. See also the M43 mortar.

In service with the Soviet, all other Warsaw Pact, Albanian and Chinese forces.

M43 heavy mortar

USSR

Calibre	120 mm
Barrel length	185 cm
Baseplate dimensions	100 cm diameter
Weight in firing position	170 kg
Carriage weight	340 kg
Range	5700 m (minimum 400 m)
Rate of fire	12–15 rpm
Operation	Manual or automatic
Sight	MP-41/MP-42 collimating sight
Ammunition	HE (15·4 kg), smoke (16 kg), incendiary (16·7 kg)
Crew	5 or 6
Manufacturer	State factories

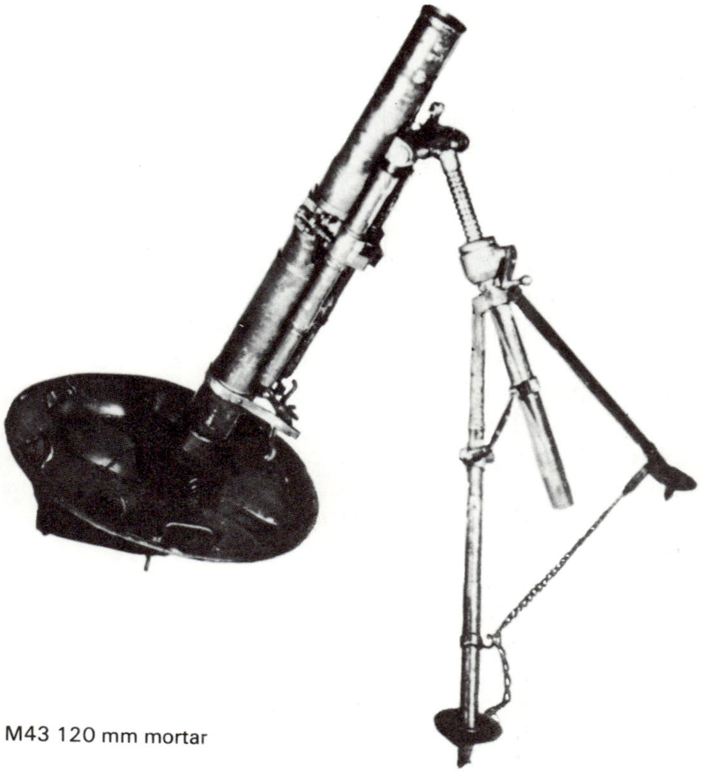

M43 120 mm mortar

The 120 mm mortar M43 has virtually replaced the M38 as the regimental mortar of the Soviet infantry and is found, with the M38, in the other countries listed, being manufactured also in communist China as the 120 mm mortar type 55.

Whilst the ballistic and performance details and methods of handling are the same for both models, the outward appearance differs slightly: the shock absorber cylinders of the M43 are much longer than those of the M38, and in the later model the elevating and traversing gear is rather more sophisticated.

The outer casing of the HE ammunition can be made of either wrought or cast iron, the latter being more effective against personnel but being slightly reduced in maximum range.

In service with the Soviet, all other Warsaw Pact, Albanian and Chinese forces.

M1 medium mortar

United States

Calibre	81 mm
Barrel length	127 cm
Barrel weight	20·2 kg
Bipod weight	19·2 kg
Baseplate weight	20·4 kg
Weight in firing position	60·1 kg
Range	3000 m
Rate of fire	20–30 rpm
Operation	Automatic
Sight	Optical cross-levelling
Ammunition	HE, HE light (3·2 kg), smoke, illuminating
Crew	3
Manufacturer	Department of the United States Army

Although at first sight there appears to be nothing to distinguish the M1 mortar from many other mortars of this size, there are two singular features incorporated in the rectangular baseplate—the base of the tube is received by any of a row of three ball-sockets (a feature shared by the French M44 mortar) and the single carrying-handle has a leather-covered grip. Otherwise the M1 is a standard smooth-bore weapon, firing HE (light and heavy), illuminating and smoke (WP and FS) ammunition. The maximum range shown can only be achieved by the light (3·2 kg) HE rounds, all of the other bombs being capable of around 2300 m.

In service with the United States, Brazilian, certain European, Japanese, Vietnamese, Yugoslav and certain Middle and Far Eastern forces, but obsolescent.

81 mm M1 mortar

M8 light mortar

Yugoslavia

Calibre	50 mm
Barrel length	56 cm
Baseplate spade length	51 cm
Weight in firing position	7·3 kg
Range	480 m (minimum 135 m)
Rate of fire	25–30 rpm
Operation	Manual
Sight	White line on barrel
Ammunition	HE, smoke, illuminating, signal (all 0·9 kg)
Crew	1–2
Manufacturer	State factories

The 50 mm Yugoslav mortar is a slightly modified version of the British World War II 2 in mortar Mk 8. It is the smallest of the series of very small, light mortars developed by the Yugoslavs and derived mainly from Western European and American mortars. It is particularly useful for guerrilla warfare in mountainous regions, such as Yugoslavia. This mortar has no baseplate and comprises instead a spade, hinged to the breech, and a barrel on which the sight consists merely of a line painted along it. The M8 is necessarily trigger fired as the round is not sufficiently heavy to be fired by the drop method. The very short range is due to the shortness of the barrel, and it can be fired from the hip.

In service with the Yugoslav forces.

M57 light mortar

Yugoslavia

Calibre	60 mm
Barrel length	70 cm
Barrel weight	5·5 kg
Bipod weight	4·5 kg
Baseplate weight	8·8 kg
Weight in firing position	19·7 kg
Range	1690 m (minimum 74 m)
Rate of fire	25–30 rpm
Operation	Automatic
Sight	Optical cross levelling
Ammunition	HE (1·4 kg)
Crew	2
Manufacturer	State factories

The 60 mm mortar M57, copied from the United States World War II M2 mortar of the same calibre, is the next to smallest of the light mortars built by the Yugoslavs for use in guerrilla warfare in their own mountainous terrain. With a total weight of only 19·7 kg, the mortar can be easily and quickly manoeuvred and transported by one man while his partner carries the ammunition. Great flexibility is given the M57 by its speed of manoeuvre, fast rate of fire, and wide variety of range—the minimum range of 74 m could even be dangerous to the firer.

In service with the Yugoslav forces.

60 mm M57 light mortar

M57 60 mm mortar bombs

M31 medium mortar

Yugoslavia

Calibre	81 mm
Barrel length	131 cm
Barrel weight	21 kg
Bipod weight	19 kg
Baseplate weight	20 kg
Weight in firing position	61·5 kg
Range	4100 m (minimum 85 m)
Rate of fire	25 rpm
Operation	Automatic
Sight	Optical cross levelling

Ammunition	Heavy HE (4·2 kg), light HE, smoke and illuminating (3·2 kg)
Crew	3
Manufacturer	State factories

The Yugoslav M31 81 mm mortar is the older of the two mortars of this calibre based on French models. The M31 has now been largely replaced by the 60 mm M68 mortar, probably also of French origin but, having been produced extensively for export, is likely to be found in Near Eastern and African countries, as well as being held in reserve in the parent country. This mortar closely resembles the United States 81 mm mortar M1, and is distinguished by a rectangular baseplate and a comparatively long, thin barrel, held by the bipod clamp at almost half-way. This mortar can be manpacked by the three crew members or otherwise transported in a light vehicle.

In service with the Yugoslav forces.

81 mm M31 medium mortar

M68 medium mortar

Yugoslavia

Calibre	81 mm
Barrel length	164 cm
Barrel weight	16 kg
Bipod weight	13 kg
Weight in firing position	41·5 kg
Range	4800 m (minimum 90 m)
Rate of fire	25 rpm
Operation	Automatic
Sight	Optical cross levelling
Ammunition	Heavy HE (4·2 kg), light HE, smoke and illuminating (3·2 kg)
Crew	3
Manufacturer	State factories

The fragile-looking 81 mm mortar M68, being an almost identical copy of the French standard mortar MO-81-61L of the same calibre, is a very much improved version of the 81 mm mortar M31, and was also produced with export in mind. In contrast to its predecessor, an increase in range of 700 m has been coupled with a decrease in overall weight of 20 kg by replacing the old rectangular baseplate with a smaller triangular baseplate, reducing the bipod to absolute minimum dimensions and producing the barrel (now even longer than before) from a light alloy. Mobility when used in a manpack role is thus considerably enhanced.

In service with the Yugoslav and some Near Eastern and African forces.

81 mm M68 medium mortar

BB1 medium mortar

Yugoslavia

Calibre	82 mm
Barrel length	120 cm
Barrel weight	21 kg
Bipod weight	20 kg
Weight in firing position	62 kg
Range	4200 m (minimum 82 m)
Rate of fire	20–25 rpm
Operation	Automatic
Sight	Optical cross levelling
Ammunition	HE (3·5 kg)
Crew	5
Manufacturer	State factories

Yugoslavia has so far produced only one 82 mm mortar, but its name is not yet known. It is most likely that the design of the mortar is based on the Soviet 82 mm mortar M37 (new type), the only visible differences being the changed shape of the baseplate (rectangular instead of round) and the connection of the bipod centre spindle to the bipod leg below the right (instead of to the opposite) leg. The Yugoslav 82 mm mortar is slightly heavier than the Soviet version and fires a slightly heavier round over 1000 m further (although a special propulsion charge is required to attain maximum range). 81 mm ammunition can also be used in this weapon. As with other Yugoslav mortars up to this size, the 82 mm can be man-packed by the crew.

In service with the Yugoslav forces.

UBM52 heavy mortar

Yugoslavia

Calibre	120 mm
Barrel length	129 cm
Weight in firing position	400 kg
Carriage weight	420 kg
Range	6000 m (minimum 195 m)
Rate of fire	25 rpm
Operation	Manual or automatic
Sight	Optical cross levelling
Ammunition	Heavy HE (15·9 kg), light HE, smoke and illuminating (12·2 kg)
Crew	5
Manufacturer	State factories

UBM52 (note the recoil system)

Although the Yugoslav 120 mm mortar (UBM52) can fire the same ammunition as the Soviet 120 mm mortars, the resemblance ends there. The UBM52 was designed as a heavy mobile weapon for use both in normal operations and in mountainous terrain. It is thus semi-permanently mounted on its own two-wheeled carriage for firing, but can be taken apart for pack transport by five mules. It also possesses a muzzle attachment to permit towing by either two mules or a vehicle. The cleverly designed spiked rectangular baseplate and hydroelastic recoil system allow the mortar to be halted and fired almost immediately, no ground preparation for the baseplate being necessary. In spite of its shorter barrel, the UBM52 can outshoot the Soviet M43. If the UBM52 uses its light HE ammunition, it has an extra 300 m range and a far higher rate of fire.

In service with the Yugoslav forces.

Grenades

Grenades were first used in the 15th Century and were so relatively unreliable that specially selected and trained troops were used to handle them. Hence the formation of the grenadier regiments. But we must jump a few years, for a convenient point in history at which to begin the story of the development of the grenade is the outbreak of World War I in August 1914. At this time, the British forces had only one official grenade. This was the Hand Grenade No 1 which had been introduced into the British forces as a result of lessons learned at Port Arthur during the Russo-Japanese War. In August 1914, the grenade was in very short supply and production in quantity was not achieved until the middle of 1915 due to the complication of its manufacture and particularly of its detonator. This absurd shortage led to the production in the field of three types of improvised grenade, which was given considerable impetus on 27 December 1914 when the Germans first threw grenades into the British trenches.

The first improvised type was called the Jam-Tin bomb which was made by infantry from jam or milk tins filled with ammonal, or gun-cotton obtained from the Royal Engineers. The grenade was fused by a short length of Bickford fuse which had to be lit before throwing. The second type consisted of glass bottles filled with Jelliteer JL, a lethal mixture of prussic acid, chloroform and triacelyl cellulose, sealed with a rubber stopper through which a fuse was inserted. The third type consisted of a tin, filled with explosive and shrapnel balls attached by two tapes to a wooden throwing handle. The time fuse protruded from the centre of the lid. The manufacture of these three types of grenade, and possibly others, was officially forbidden in December 1915, when quantities of the No 1 Grenade became more readily available. It is reported that there was much relief at this.

The No 1 Grenade consisted of a glass cylindrical body encircled by a narrow cast-iron ring, serrated to break into 16 fragments. The body was mounted on a wooden block to which a can handle was attached. Four webbing streamers were nailed to the top of the handle. This grenade had a number of safety devices, including a glass or aluminium removable cap upon which were stamped the words 'Fire', 'Travel' and 'Remove'. It also had a firing pin held in position by a whipcord bracket and a leather tag. Inevitably, this grenade had to be handled with some respect for it was filled with 3 oz of explosive and detonated upon impact—if not before.

Later, Hand Grenade No 2 (the Mexican or Tonite Hales grenade) and the Rifle Grenade No 2 percussion were introduced. Over 12½ million of these were produced during the war.

Finally, in the spring of 1915, the No 5 Grenade (or Mills Hand Grenade), which was a time-fused grenade, was introduced and over 33 million of them were issued before it was superseded by No 23 Grenade. This grenade was very similar to the 36 Grenade which has just become obsolescent in the British forces. It was probably one of the most famous series of hand grenades ever to be produced.

It is interesting to note, perhaps, that at the outbreak of the war, only the Royal Engineers (and not the Grenadiers) so far as the British Army was concerned, were trained in the use of these new grenades. However, trench warfare forced a change, although the training of infantry as bombers did not start seriously until May 1915. During World War II, a very great number of grenades was manufactured. Many had considerable similarity to the grenades of World War I, but many did not. In addition to the more conventional HE-filled grenades, there were phosphorus, incendiary, sticky and gas grenades. Considerable use was made of grenades fired from rifles, and, indeed, considerable use was made of grenades during training. Battle inoculation, as it was called, became a very important aspect of training during the period of the phoney war and just after. Realism, quite rightly, was considered to be an essential part of training. For this reason, and so far as the British forces were concerned, the 69 Grenade was a training aid; nowadays it would be considered to be an offensive hand grenade.

Since World War II great strides have been taken in three main areas; first, the rifle grenade. This had become a thoroughly sophisticated weapon culminating in the ability to use live ball ammunition as a means of projecting it. Clearly, this has some advantage for it avoids the manufacture, and the need to carry, special ballistite, or blank ammunition.

The second area is one in which multi-purpose, or polyvalent, grenades have been developed. These mainly consist of four or five grenade parts, namely the fuse and its cover, the body, a pre-fragmented metal sleeve and the tail assembly. Together with a clever fuse-setting system, it is possible to produce three main configurations—offensive hand, defensive hand and rifle grenades. To these can be added an 'impact only', 'delay only' or 'impact and delay' fuse.

So far as hand grenades are concerned, therefore, there are two basic types: 'offensive' and 'defensive'. The latter are designed to produce a large number of fragments of all sorts and thus to cause maximum damage. The former (offensive) grenades are designed for use by assaulting infantry who can continue to move forward having thrown their grenades, which produce a good blast effect and daze the defender without affecting the attacker much, if at all.

DT21 offensive/defensive grenade (Spain)

RPG43 anti-tank grenade (USSR)

RG34 offensive hand grenade

Czechoslovakia

Dimensions	Diameter 6·5 cm Height 7·4 cm
Weight	0·3 kg
Body material	Sheet steel
Charge	HE (100 gm)
Arming	Pin, spring and striker
Activation	Impact
Effect	Fragments out to 15 m or 25 m
Safety	Manual pin
Manufacturer	State factories

The Czech RG34 hand grenade is an offensive anti-personnel weapon which can, when necessary for defensive purposes, be fitted with a pre-fragmented cast-iron jacket over the sheet steel body. In the latter case the grenade is lethal out to a radius of 25 m, but this is reduced to 15 m without the jacket. This grenade is not fitted with a delayed-action fuse, and after the safety-pin and tape are removed the percussion-detonating mechanism (the spring-loaded striker) is activated only by impact on the grenade body. This grenade is being superseded by the RG4 from which it differs by having flatter ends, no corrugated grip, and a more primitive type of safety clip.

In service with the Czechoslovak forces.

RG4 offensive/defensive hand grenade

Czechoslovakia

Dimensions	Diameter 5·3 cm Height 8·4 cm
Weight	0·3 kg
Body material	Sheet steel
Charge	HE (105 gm)
Arming	Clip, tape and spring-loaded striker
Activation	Impact
Effect	Fragments out to 15 m or 25 m
Safety	Clip and tape
Manufacturer	State factories

The Czech RG4 hand grenade, an offensive anti-personnel weapon, is an up-dated version of the RG34, which it is intended to supersede. Like the RG34, this grenade can, when required for defensive purposes, be quickly fitted with a cast iron pre-fragmented jacket over the sheet-steel body so as to increase its effective range from 15 m to 25 m. The grenade has been redesigned to produce a more streamlined shape with a corrugated grip to increase throwing range, whilst at the same time slightly reducing overall weight, yet increasing the HE charge. The design of the safety-clip has been simplified and made more secure, but the activation principles remain the same. Explosion on impact is by means of a spring-loaded striker, after the latter is freed by release of the safety-clip and tape.

In service with the Czechoslovak forces.

S32 smoke hand grenade

Germany (DGR)

Weight	250 gm
Body material	Cardboard
Charge	Composition
Arming	Friction cap
Activation	7 sec delay ignition fuse
Effect	Black smoke up to 2 min
Safety	Cap protects ignition pellet
Manufacturer	State factories

The S32 East German smoke hand grenade is the older and smaller of two grenades of similar construction used in the Nationale Volksarmee. The contents of the light cardboard container are easily and quickly activated by twisting the small cap on the top of the grenade—a small abrasive pad inside the cap ignites an ignition pellet which in turn ignites the delayed-action fuse. Care should be taken to throw the grenade without delay, as the 7 sec fuse is not always accurately measured. Instructions for use are printed on the side of the body under the designation Rauchkorper (smoke body).

In service with the DGR forces.

S53 smoke hand grenade

Germany (DGR)

Weight	300 gm
Body material	Cardboard
Charge	Composition
Arming	Friction cap
Activation	7 sec delay ignition fuse
Effect	White smoke for 30 sec
Safety	Cap protects ignition pellet
Manufacturer	State factories

The S53 East German stick smoke grenade is, apart from its handle, of similar construction to the S32 smoke grenade. The grenade is activated in a similar fashion by twisting the protective cap to ignite the ignition pellet and thus the delayed-action fuse. The filling, however, is of a completely different faster burning material producing white smoke, but has the advantage of being able to continue producing smoke, even if submerged in water. As with the S32, instructions for use are printed on the side of the cardboard cylinder beneath (in this case) the designation Nebelhandgranate (fog hand grenade). This grenade, too, should be thrown straight after igniting as, once again, the 7 sec fuse is not wholly reliable.

In service with the DGR forces.

M42 offensive hand grenade

Hungary

Dimensions	Height 19·4 cm Diameter 4·8 cm
Weight	0·3 kg
Body material	Sheet steel
Charge	HE (134 gm)
Arming	Pin, spring and striker
Activation	3–5 sec delay pyrotechnic train
Effect	Blast to 25 m
Safety	Manual pin
Manufacturer	State factories

The main reason for the retention in service of the Hungarian World War II stick grenade is probably the interesting design feature of threading at both ends of the grenade body which allows several grenades, after removal of the handle, to be screwed together to form a demolition charge. Otherwise the weapon is a normal offensive anti-personnel grenade designed for clearing bunkers and trenches, with a dangerous and unreliable delayed-action fuse of the percussion-operated type, activated simply by extraction of the safety-pin releasing the striker spring.

In service with the Hungarian forces.

F1/N60 defensive rifle grenade

Poland

Dimensions	Diameter 55 mm Height 270 mm
Weight	630 gm
Body material	Cast iron
Charge	HE (45 gm)
Fuse	Impact
Effect	Fragments out to 20 m
Manufacturer	State factories

F1/N60 rifle grenade

The Polish Army uses two rifle grenades—the F1/N60 anti-personnel fragmentation grenade and the PGN-60 anti-tank grenade—a type of weapon employed by no other communist country except Yugoslavia. These grenades are fired from the PMK-DGN—an AK-47 modified with a shoulder pad strapped to the butt, a large tangent leaf rearsight graduated 100–240 m, and a spigot-type muzzle attachment. The grenade is also a Polish modification of a Soviet weapon—in this case the F1 fragmentation hand grenade, which it largely resembles, but is fitted with a simple stabilising boom, has no safety-pin ring, and uses an impact fuse in place of the F1's delay fuse. The pre-fragmentated body enables defending troops to increase the effectiveness of their fire against an attacker by launching anti-personnel grenades very accurately and rapidly to a distance far exceeding the normal grenade hand throw. A special small magazine can hold ten launching cartridges, and is made with the forward side blocked so that normal cartridges cannot be loaded in error.

In service with the Polish forces.

PGN-60 HEAT rifle grenade

Poland

Dimensions	Diameter 67·5 mm Height 405 mm
Weight	580 gm
Body material	Steel
Charge	HEAT (218 gm)
Range	90 m
Fuse	Impact
Effect	Concentrated blast penetrates up to 100 mm of armour plate
Manufacturer	State factories

The PGN-60 HEAT rifle grenade is of wholly Polish design, unlike the other of the two Polish rifle grenades, the F1/N60 anti-personnel grenade. Of all communist countries, only Poland and Yugoslavia employ this type of grenade. Like the F1/N60, the PGN-60 is launched using the Polish modification of the Soviet AK-47—the PMK-DGN, fitted with shoulder pad, tangent leaf rearsight graduated 100 m–240 m, a spigot-type muzzle attachment, and 10 round magazine. The sights, however, appear to have been designed basically for the anti-personnel F1/N60 grenade, as the effective anti-tank range of the PGN-60 is only about 90 m, maximum penetration capability of the shaped-charge warhead being in the region of 100 mm. Stabilisation in flight is ensured by an extended tail-boom fitted with narrow fixed fins.

In service with the Polish forces.

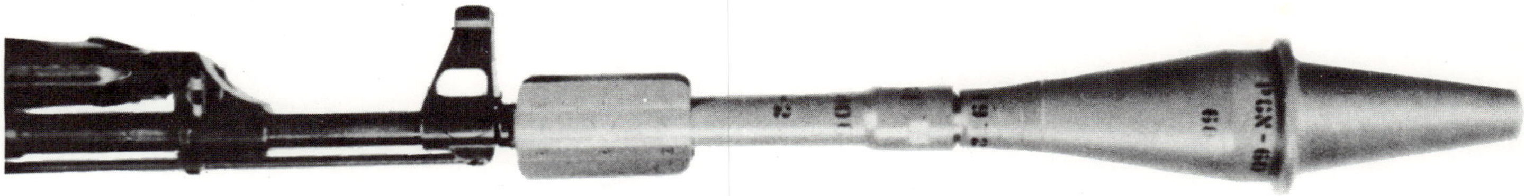

PGN-60 rifle grenade

RDG-2 (and RDG-1) smoke hand grenade

USSR

Dimensions	Diameter 4·5 cm Height 25 cm
Weight	500 gm
Body material	Waxed cardboard
Charge	Smoke composition
Arming	Friction cap
Fuse	5 sec delay pyrotechnic train
Effect	White smoke for 100 sec
Manufacturer	State factories

The Soviet RDG-2 is a long, narrow cylinder of waxed cardboard, activated by the normal friction cap and delay-fuse principle. The white smoke cloud produced covers an area of about 20 m by 10 m after burning for some $1\frac{1}{2}$ minutes.

RDG-1 is a similar grenade of waxed cardboard, but has a wooden handle to facilitate throwing and two slightly different models produce either white or black smoke. The activation principle is the same as that of RDG-2, but this grenade has one further advantage. Owing to its wooden handle, it does not sink, and will continue to burn and produce smoke in water.

In service with the Soviet and other Warsaw Pact forces.

RGD5 defensive hand grenade

USSR

Dimensions	Diameter 5·6 cm Height 11·4 cm
Weight	310 gm
Body material	Sheet steel
Charge	TNT (110 gm)
Arming	Pin, lever and striker
Fuse	3–4 sec delay pyrotechnic train
Effect	Fragments out to 25 m
Safety	Manual pin
Manufacturer	State factories

The latest development in Soviet anti-personnel fragmentation grenades is the defensive RGD5, which is only just over half the weight of the older F1 but retains a similar effective range and can naturally be thrown much further. The sheet steel body is smooth, oval and stamped in two pieces with a raised lip round the centre, with, however, little change in the method of arming and activation. The 3–4 sec delay fuse is activated by the striker being released when the safety lever is allowed to pivot on the end of the long fuse body. The fragmentation jacket is fitted inside the grenade body, being serrated in the case of the Soviet model and smooth-surfaced in the Chinese version, which is called the M32. This grenade, like most Soviet hand grenades, uses the UZRG time fuse.

In service with the Soviet, other Warsaw Pact, Chinese and Vietnamese forces.

RGD5 hand grenade

RKG3/3M/3T anti-tank hand grenade

USSR

Dimensions	Diameter 5·5 cm Height 36·2 cm
Weight	1070 gm
Body material	Sheet steel
Charge	HEAT (565 gm)
Arming	Pin, inertial striker
Fuse	Impact
Effect	Shaped charge, concentrated blast
Safety	Manual pin
Packing	Body, fuse and handle separate
Manufacturer	State factories

Most Warsaw Pact countries now use the RKG3 or its modernised version, the RKG3M, as the standard infantry anti-tank hand grenade to supplement the RPG7V anti-tank grenade launcher, and both systems have been sent to Vietnam. The RKG3 series grenades are much heavier and longer than the Soviet anti-personnel grenades and are transported in the three separate parts—the cylindrical body containing 567 gm of HE shaped-charge filler, the handle containing the safety-pin and a drogue parachute assembly, and the small detonator. After assembly, the grenade is operated as follows: the handle is held in one hand while the other hand removes the safety ring and pin—this action frees both the detonator and the handle cap retaining the parachute. The grenade is now thrown and the parachute retaining cap releases the parachute which ensures that the shaped-charge warhead strikes the target at the correct impact angle, whereupon the firing pin, located in the handle, is forced down to strike the detonator. Owing to its weight, the RKG3 can only be thrown about 20 m, but it can penetrate up to 120 mm of armour plate. The only difference between RKG3 and RKG3M is that the latter has a copper charge-lining. The Chinese copy is called RKG3T and is of exactly the same design as the Soviet model.

In service with the Soviet, other Warsaw Pact and Vietnamese forces.

RKG3 hand grenade

F1 defensive hand grenade

USSR

Dimensions	Diameter 5·5 cm Height 12·4 cm
Weight	690 gm
Body material	Cast iron
Charge	TNT (60 gm)
Arming	Pin, lever and striker
Fuse	3–4 sec delay pyrotechnic train
Effect	Fragments out to 20 m
Safety	Manual pin
Manufacturer	State factories

Like the RG42, the F1 anti-personnel fragmentation defensive grenade was first produced in World War II and has since been used (and still is) by most communist countries. Although somewhat similar in operation and appearance to the British Mills grenade, having an oval heavily-serrated body and a long side-mounted safety lever, the F1 is, however, loaded by inserting the detonator into the top after unscrewing the elongated fuse body, which is then replaced. As with the RG42, the spring-loaded striker activates the fuse when released by the safety lever being now allowed to pivot when the grenade leaves the throwing hand. Whilst the danger zone is given as 20 m radius, it is not unknown for fragments to cause injury at 100 m. Poland uses a modification of the F1 (with an impact fuse) as the F1/N60 anti-personnel rifle grenade. The Chinese model is called the Type 1/M33, and is an exact copy, apart from markings. The grenades can be thrown about 45 m.

The fuse of the F1 can be either the standard UZRG or the F1, the latter burning 1/5th sec longer. Owing to the large fragmentation area, the grenade should be thrown from behind cover.

F1 hand grenade

In service with the Chinese, Soviet, other Warsaw Pact and Vietnamese forces.

RG42 defensive hand grenade

USSR

Dimensions	Diameter 5·4 cm Height 12·7 cm
Weight	420 gm
Body material	Sheet steel
Charge	TNT (118 gm)
Arming	Pin, lever and striker
Fuse	3–4 sec pyrotechnic train
Effect	Fragments out to 20 m
Safety	Manual pin
Manufacturer	State factories

The RG42 anti-personnel stick grenade dates back to World War II, but can still be found in many communist countries. Although its outward appearance is that of an offensive blast grenade, inside the sheet-metal cylinder body is a light serrated metal liner, which is fragmented on detonation. The weapon is not thrown by its handle, but is operated in a similar fashion to the British Mills grenade. When the grenade is thrown, the lever is released and it springs from the cylinder, allowing the detonator, housed in the handle, to strike and activate the 3–4 sec delay fuse, the UZRG standard fuse employed in most Soviet hand grenades. The grenade can be thrown about 40 m.

In service with the Chinese, Soviet, other Warsaw Pact and Vietnamese forces.

RG42 hand grenade

M69 offensive hand grenade

Yugoslavia

Dimensions	Height (without cap) 11 cm Diameter (maximum) 5·5 cm
Weight	600 gm
Charge	TNT
Body material	Cast iron
Fuse	4–5 secs delay
Effect	20–25 m
Manufacturer	State factories

The Yugoslav M52R and M69 anti-personnel offensive hand grenades closely resemble each other, the only real difference being that the older M52R is fitted with an impact fuse and the M69 has been given a safer delay fuse. Both are of an oval shape with a large fuse body extending from one end which is unscrewed to allow insertion of the detonator. Of similar size, these two grenades are very small for their actual weight (almost twice that of most other hand grenades used by communist countries).

In service with the Yugoslav forces.

M69 hand grenade

M60 anti-personnel and HEAT rifle grenades

Yugoslavia

M60 ANTI-PERSONNEL RIFLE GRENADE

Dimensions	Height 30·7 cm Diameter 3 cm
Weight	520 gm
Fins	4 white plastic
Range	400 m maximum
Charge	HE—69 gm Redex TNT
Body material	Steel
Fuse	Impact
Effect	Shrapnel to 20 m
Armed	By unscrewing nose cap
Manufacturer	State factories

M60 HEAT RIFLE GRENADE

Dimensions	Height 39 cm Diameter 6 cm
Weight	602 gm
Fins	6 white plastic
Range	150 m maximum
Charge	Shaped—235 gm Redex TNT
Body material	Steel
Fuse	Impact
Effect	Penetrates armour up to 200 m (19 cm)
Armed	By unscrewing ring on tail
Manufacturer	State factories

M60 HEAT markings

Rifle grenades are not common in Warsaw Pact armies and are in service with only the Yugoslavia and Polish forces. These well-made and very efficient grenades have a 22 mm interior boom diameter and can be fired from NATO weapons with a similar diameter grenade launcher attachment (eg FN, FAL, G3, etc). The ballistite cartridge is contained in a plastic plug which inserts into the tail of the grenade. So far as these grenades are concerned, they can only be fired from the Yugoslav M70/64 and M59/66 rifles and from some specially modified Mauser Kar 98 in 7·92 mm calibre.

In service with Polish and Yugoslav forces and Rhodesian guerrillas.

M60 anti-personnel (right) and HEAT (left) rifle grenades

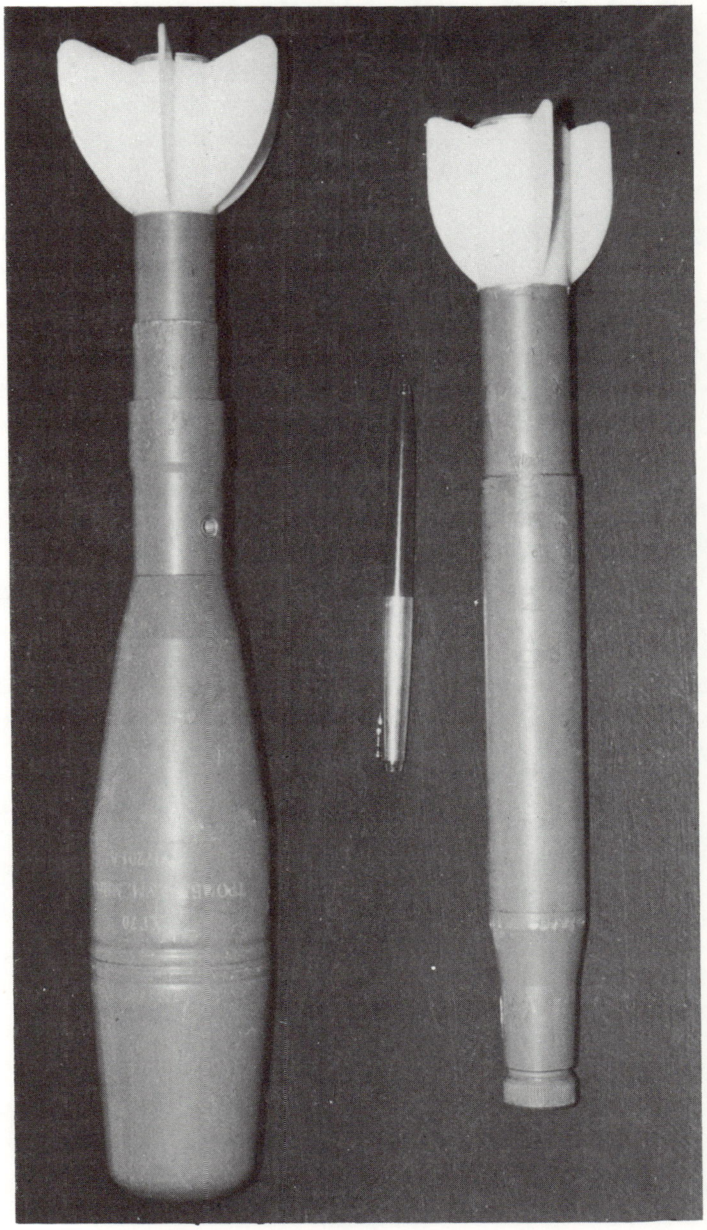

M60 fragmentation rifle grenade

Yugoslavia

Dimensions	Diameter 3 cm Height 31 cm
Weight	520 gm
Body material	Sheet steel
Charge	HE (67 gm)
Range	400 m
Fuse	Impact
Effect	Fragments out to 50 m
Manufacturer	State factories

The Yugoslavs and the Poles are the only communist nations to employ rifle grenades. Until a short time ago Yugoslavia imported rifle grenades from Belgium, but is now producing her own anti-personnel and anti-tank rifle grenades, both of which have a very slim, streamlined shape. The anti-personnel fragmentation grenade has an excellent range of up to 400 m, being fired by the spigot-type launcher of the M59/66 Yugoslav modification of the SKS carbine, or from the older M48 rifle. Although fitted with rounded-tail stabilising fins similar to those on the anti-tank rifle grenade, this missile is smaller, lighter and of almost equal diameter for nearly the whole of its length. It is cylindrical and has a flat, slightly tapered nose, containing the impact fuse, whilst the HE filling occupies the central third of the grenade.

In service with the Yugoslav forces.

M62 smoke rifle grenade

Yugoslavia

Dimensions	Diameter 4 cm Height 33 cm
Weight	480 gm
Range	465 m with 7·9 mm rifle
Fuse	Base detonating with 7–8 sec time delay
Manufacturer	State factories

M62 illuminating rifle grenade

Yugoslavia

Dimensions	Diameter 4 cm Height 33 cm
Weight	450 gm
Range	240 m with 7·9 mm rifle
Fuse	Burning time 30 sec
Manufacturer	State factories

The illuminating and the smoke rifle grenades, together with the anti-personnel and the anti-tank rifle grenades are very recent developments in Yugoslavia. All are long, streamlined missiles, using similar rounded tailfin assemblies, differing in appearance only slightly. Both the smoke and illuminating grenades are torpedo-shaped, the front half slightly thicker than the rear half, and have the safety-pin, with a ring attached, inserted just below half-way. Whilst the nose of the smoke grenade is smooth and rounded, that of the illuminating grenade is semi-elliptical and is encircled by a milled band. As with the anti-personnel and anti-tank grenades, these other two missiles can be fired from the spigot-type launcher of the M48 rifle and the M59/66 Yugoslav modification of the Soviet SKS carbine.

In service with the Yugoslav forces.

Flamethrowers

Fire has been employed as an instrument of warfare for hundreds of years, but the modern flamethrower was Richard Fiedler's conception. In 1901, this Berlin engineer offered two models for the German army to try. The smaller *flammenwerfer* was sufficiently light to be carried by one man and used gas pressure to eject a stream of flaming oil for a distance of about 20 m; while the bigger version had a range of 40 m and held enough fuel for 40 sec continuous firing and needed a team to handle it.

The principle on which the *flammenwerfer* worked was simple; a valve released gas in the lower chamber which pushed liquid in the upper chamber into a rubber pipe. The fluid was held in check by two other valves before reaching the ignition device. this device comprised a small tube containing a spring, a detonator, gun cotton and a paraffin wick. At a point when the gas forced the liquid against the spring, the wick ignited and a jet of flame projected for 20 m, or more. The duration of the flame was approximately two minutes.

The equipment was subsequently issued to the 23, 24, 25 Pioneer Battalions and in 1912, the first *Flammenwerfer* regiment was formed consisting of 12 flamethrower companies and named the 3rd Guard Pioneer Regiment. The German flamethrowers were a success at the Battle of Hooge in 1915 because the British and German trenches were very close together. Tactically, flamethrowers were worth deploying only if the forward lines of the opposing forces were no further apart than the maximum range of the equipment—and herein lay its weakness and the cause of frustration for the operators of the equipment.

In the 1939–45 War, with the onset of mechanisation and mobility, it was inevitable that the flamethrower would be substantially modernised. The small man-portable infantry flamethrower was still used, but some larger equipments were now fitted to tanks. The most effective and imaginative of all the flamethrowers was the British Crocodile mounted on the Mark VII Churchill tank. The flamethrower could only project its hot shot up to 110 m and usually did not open fire until about 75 m. A 6 ton trailer towed behind the tank carried 1800 l of napalm and the tow-bar carried the linkage that supplied the fuel to the gunner's position. Impetus was applied by nitrogen gas from five cylinders at an initial pressure of 210 kg/sq cm reduced to a working pressure of 20 kg/sq cm. The flame fuel could be shot at 181 m/sec in a solid 'rod' ignited by a spark and petrol vapour system. It was possible for a good gunner to despatch this 'rod' through the slit in a pill box at extreme range.

In future wars, on terrain suitable for mobile warfare, the use of flamethrowers is likely to be limited, for tanks towing such clumsy equipment will be very vulnerable. On the other hand, in urban areas, smaller flamethrowers, or fire-raising devices of some sort, will be extremely valuable to attacker and defender alike, for instance a launcher which fires a red phosphorous grenade several hundred metres would not be impossible to produce, and could be as effective as it is unpleasant. Such weapons will remain useful for some time to come and it is a pity that development of them in the forces of the Western World has been so reduced.

ROKS-2 manpack flamethrower

USSR

Capacity	10·9 kg (10·4 l)
Range	35 m
Manufacturer	State factories

The ROKS–2 manpack flamethrower is the oldest of the three weapons of this portable type used by Warsaw Pact countries. It is not likely to be met with in the Soviet Army, but will be in some Warsaw Pact forces and it is possible that communist China and North Vietnam may still retain the ROKS–2 in service. Cunningly designed as a surprise weapon, the ROKS–2 assembly looks very much like a rifle and haversack.

The assembly consists of the rifle-like launcher, connected by a flexible hose from the base of the 'breech' to the bottom of the pack containing the flat, box-like, sheet-steel fuel tank, with a thin flexible air-pipe feeding the fuel-pipe from the compressed air bottle, secured also to the base of the fuel tank. In order to effect minor adjustments and repairs to the fuel and air supply system, a small set of tools is also provided, contained in a small rectangular canvas pocket attached to the front of the fuel tank. The tank is filled through an inlet with a one-way valve on top of the tank. For carriage, a double sling is fitted to the launcher and shoulder-straps to the fuel container. The ignition cartridges, activated by the trigger, are to be found in the cylindrical muzzle attachment.

In service with some Warsaw Pact forces.

ROKS-3 manpack flamethrower

USSR

Weight	23·5 kg (filled)
Capacity	8·8 kg (8·2 l)
Range	35 m
Ignition charges	3
Firing time	5 sec per charge
Manufacturer	State factories

The Soviet-made ROKS–3 portable flamethrower is a slightly more modern version of the ROKS–2, but is likewise being phased out and is being replaced by the manpack LPO–50. In this case, however, there is little attempt to deceive the enemy as, although the launcher still resembles a rifle, the fuel tank is completely undisguised. The ROKS–3 comprises a rifle-like launcher equipped with a cylindrical 10 round ignition cartridge drum fitted round the muzzle, a flexible hose connecting the launcher 'breech' to the base of the cylindrical fuel tank, and a compressed air cylinder secured to the side of the fuel tank. The fuel tank inlet is located centrally on top of the tank. With a full tank, the whole assembly weighs 23·5 kg and, as with ROKS–2, is carried by shoulder-straps on the back of the operator. To operate, one pull on the trigger opens the fuel hose valve and a second pull activates one of the 10 ignition cartridges. It is claimed that sufficient fuel is carried to permit a 5 sec burst with each cartridge. The maximum range of 35 m can only be attained by using thickened fuel—otherwise, the range is reduced to 15 m. A burst is completed when the trigger is released, automatically closing the spring-loaded hose valve. As with the ROKS–2, the launcher is provided with a rifle-type sling for carriage when not in use.

In service with the Chinese, some Warsaw Pact and Vietnamese forces.

Sketch of ROKS-3 manpack flamethrower

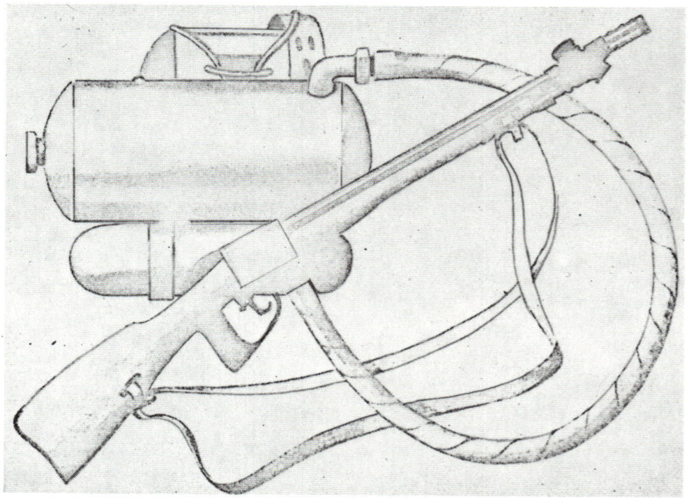

LPO-50 manpack flamethrower

USSR

Weight	25 kg (filled)
Capacity	10·6 kg (10 l)
Range	70 m
Ignition charges	3
Firing time	3 sec per charge
Manufacturer	State factories

The LPO–50 is replacing the old ROKS–2 and ROKS–3

portable flamethrowers in the Soviet Army and most Warsaw Pact forces. Compared to its predecessors, the LPO–50, with a similar weight, carries more fuel and boasts over twice the range. This weapon consists also of a rifle-like launcher fitted with a folding bipod, a pistol-grip and carrying-sling, a flexible hose connecting the metal butt casing to the triple tank assembly, and the tank assembly itself. The three tanks are fixed side by side on a manifold tubular base which feeds fuel directly into the fuel hose, one-way valves being located in the base of each tank to prevent any return flow. No compressed air bottle is required, as each tank carries on top next to the safety valve a pressurising charge mounted in the cap of the fuel inlet. Shoulder-straps are fitted to allow it to be carried on the back of the operator, who is also issued with a protective face-mask. The whole assembly weighs 23 kg, but the tank-assembly can be reduced to two tanks, or one, if damage or some other reason makes this necessary. Normal trigger operation releases the fuel and activates the electrical ignition system contained in the nozzle. In addition, however, the trigger action electrically activates the pressurising charge in the filler cap. The maximum range of 70 m can only be attained if thickened fuel is used, otherwise 20 m is the maximum range.

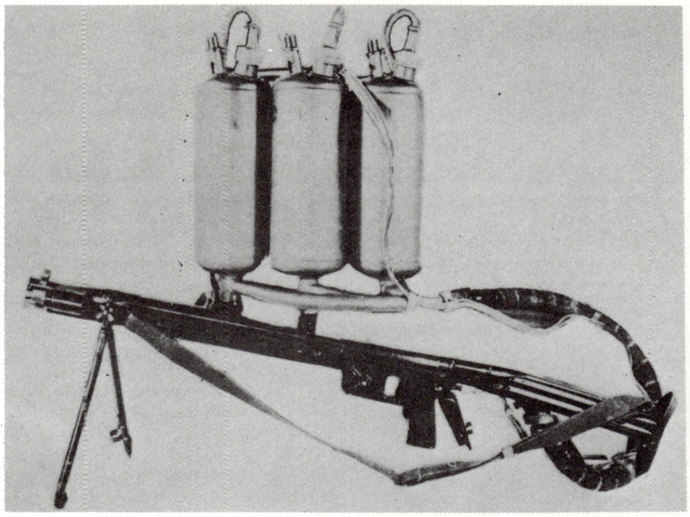

LPO-50 manpack flamethrower

In service with the Soviet, most Warsaw Pact and the Vietnamese forces.

Mines

Mining in the true military sense began soon after gunpowder ceased to be regarded as a combat aid, in the sense that Chinese crackers and Greek fire were combat aids, and it became in the 14th century a weapon of war. The destructive effects of explosives were soon recognised, and it was not long before the tacticians of the Middle Ages realised the potential, given them by explosives, for breaching fortifications. The research establishments in those days soon discovered that the most effective way to destroy a target was to place charges beneath it. It was assumed, rightly, that this method would be quicker and surer than laying siege, or marching round and round a city blowing trumpets waiting for the walls to fall, like Joshua. It was also likely in theory at any rate to be less harmful to the attackers than storming the walls or exposing themselves to boiling oil by operating battering rams to achieve this aim, as the Crusaders and others quickly found. However, the same problems of approach, concealment and the correct placing of the charges remained. All this called for special skills, intelligence and some luck, so units of sappers and miners were raised and trained for the purpose.

Gunpowder remained the only explosive available for such operations until late into the 19th century and even then little technical advance was achieved. However, considerable research and development effort was put into the manufacture of new explosives during the 1914–18 war. The static warfare on the Western Front was such that ground could only be won by disastrous frontal assault. Despite intensive artillery bombardments and heavy machine gun support, casualties were shatteringly high and, in an effort to reduce these, other means of gaining ground had to be resorted to. One such means was mining, which culminated, towards the end of the war, in the detonation of colossal quantities of explosives. This resulted in the displacement of hundreds of tons of earth and the formation of gigantic craters which literally obliterated salient features in the front lines. The explosive favoured by many mining engineers was ammonal. This was a grey powder with a power three times greater than that of gunpowder. There were, of course, complications in that ammonal was hygroscopic which made it unsuitable for any lengthy storage underground. It was also corrosive to copper, so aluminium-sheathed detonators had to be devised. Amatol was an alternative explosive without the faults of ammonal, but less efficient.

Examples of mining during World War I ranged from relatively modest charges of gunpowder (130 lb–900 lb) in 1915 in the Ypres area to an impressive charge of 1,000,000 lb made up of 23 separate charges and fired together on a single morning in 1917 to herald the British offensive into Flanders. It was during this period, due to the hygroscopic effect mentioned, and the shortage of gunpowder as well, that ammonal was packed experimentally into water-tight tins and used very successfully to set off charges.

During the 1930s, the increasing use of wheeled and tracked vehicles led to an early realisation by some, and confirmation of their views by others, such as the Generals Fuller and de Gaulle, that armoured warfare would play a more and more important part in any future war. Clearly, some form of countermeasure to combat armoured fighting vehicles was necessary. Thus, there began in earnest the escalating struggle between the anti-tank gun on the one hand and the tank's offensive capability of its gun, its armour and its speed and manoeuvrability on the other. An equally effective, but much cheaper means of defence against the tank was the use of land minefields, whereby powerful explosive devices could be deployed defensively across the front, as at sea, to stem the advance of attacking tanks by blowing off their tracks. All that was needed was a tin packed with explosive and coupled to a fusing device which could be set off by a heavy weight. The first real anti-tank mine was brought into British service in September 1935.

This new and devilish device consists of a round steel body some 18 cm in diameter and 5 cm high and was filled with baratol 10/90. The central chamber accommodated a firing pistol which consisted of a body containing a spring-loaded striker restrained by a sheer-wire. Beneath the striker, there was a cap placed alongside a cavity in the central chamber which housed four perforated CE pellets surrounding a detonator. The mine was surmounted by a circular domed steel cover, which was held off the mine by a strong spider/spring and designed to give under a pressure of some 165 kg. The mine was buried about $7\frac{1}{2}$ cm below ground level. When a heavy weight passed across it the cover was depressed until the sheer-wire parted. This allowed the spring to force the spider downwards to fire the cap. The flash from the cap fired the detonator which set off the exploder and the main filling, and up, it was hoped, went the target.

Mines of various descriptions and complexities followed from this, the first of its kind. They came in various sizes and with various devices including anti-lift devices and booby-traps. It is only latterly, perhaps, that new shapes have materialised, mainly because of the need to either accommodate new devices, reduce their chances of detection, or to enable them to be laid mechanically. Latterly, too, developments have included the use of new materials which cannot be detected electro-magnetically, and materials which have greater heat or cold, and water resistance.

In addition to these charges which would be predictable in an age of quick technological advance, mine development has taken a turn outside the conventional hole-in-the-ground concept. The French (DTAT) have developed the MAH F1 mine which, in simple terms, is a sort of missile launcher set at waist height and triggered off either by a tripwire or by a remote control. There is Astrolite, a liquid explosive, developed by the United States which can be sprayed on to the ground and is difficult to detect. Gaining favour too in some countries is the use of standard charges which can be built up sequentially as necessary into a mine of any size but not, one imagines, to the size of the mountain of ammonal set off by General Plumer at 0310 on 7 June 1917. That sort of cratering can be more practically carried out by nuclear mines, although not so simply politically.

In this section of the book, we concern ourselves mainly with the more conventional mines and not with demolition charges, although the borderline in some cases is difficult to determine. We also deal briefly with detection but make no mention of mine laying—a skill carried out usually by infantry, though with engineer support and advice. Mining is a complicatedly wasteful form of warfare but one which, as the Services' manpower becomes scarcer, will become more necessary.

No mention has been made of anti-personnel mines, many examples of which are to be found in this section. In general terms, the mines themselves are manufactured to a limited number of basic mechanical or material configurations. It is in their application, siting, triggering and effect that there are infinite variations and where human ingenuity knows few bounds, but the reader can discover some of this for himself in the pages that follow.

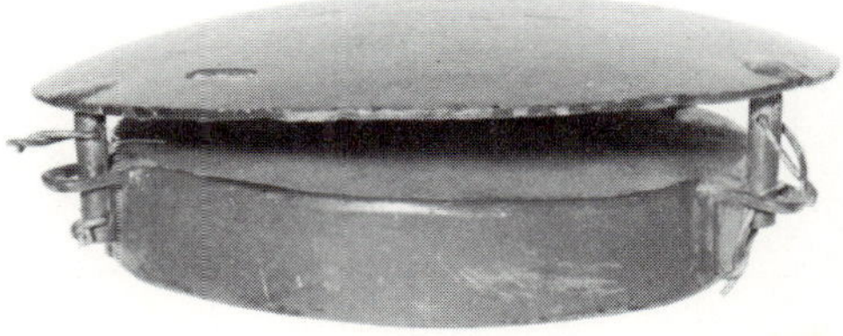

CVP-1 dual-purpose mine (Hungary)

NA-Mi-Ba anti-personnel mine

Czechoslovakia

Dimensions	Height 24·7 cm
	Diameter 19·6 cm
Weight	2·4 kg
Body material	Plastic
Charge	TNT
Manufacturer	State factories

The Czech NA-Mi-Ba mine is a tripwire-operated booby trap. It consists of a TNT charge activated by a chemical fuse, spring striker and detonator housed in a corrugated-plastic cylinder. On top of the cylinder, a lever is hinged in the centre in such a manner that, if it is moved by the wire attached to its outer end, the lever depresses a knob. Depression of this knob will fracture the two phials of chemicals in the fuse and their interaction will release the striker. Other methods of employing this mine are as an initiating charge for larger engineer demolition charges when normal electrical ignition is not available and again using the tripwire to activate anti-tank mines not detonated by a tank's tracks. As detonation results merely from depression of the knob on top of the mine, it is obvious that the opportunities for using alternate means of depressing this switch are many and varied.

In service with the Czechoslovak forces.

PP-Mi-Sr anti-personnel mine

Czechoslovakia

Dimensions	Height of casing 12·5 cm
	Diameter 10 cm
Weight	3·2 kg
Body material	Sheet steel
Charge	TNT
Manufacturer	State factories

The Czech PP-Mi-Sr bounding mine is a short, thick steel cylinder with a rounded top rim. In the centre on top is the plastic cap on the propellant charge cylinder and, on either side of that, the two smaller screw-head caps of the filler-cap and the detonator well-plug. The explosive filler (325 gm of TNT), propellant charge and ignition mechanism are housed in an inner casing surrounded by a mass of steel-rod pieces inside the outer casing. Initial activation may be by RO-1 pull-fuse using a tripwire, or by an RO-8 pressure-fuse—these fuses operate the propellant charge which, after a 3 sec delay, causes the mine to leap upwards. Attached to the propellant charge is a 1 m long securing wire. As this becomes taut, it releases the spring-loaded striker to hit the detonator, so that the mine explodes immediately—while still in the air. The advantage gained is that the explosion is activated at waist height without the mine having to be placed in view, as is the case with a stake mine such as the PP-Mi-Sb. Although the PP-Mi-Sr is much more powerful than the stake mine, it also needs quite a strong force to activate the fuse—3·5 kg, and so is less likely to be set off by small animals. Disarming is a slightly different process, depending on the fuse type. If the fuse is the RO-8 pressure type, a pin should be placed in the safety-pin aperture; if the RO-1 pull fuse is present, the tripwire (if loose) should be severed (a tight wire may be booby-trapped). The fuse can then be extracted.

In service with the Czechoslovak and some Near Eastern forces.

PP-Mi-Sb anti-personnel mine

Czechoslovakia

Dimensions	Diameter 7·5 cm
	Length 14 cm
Weight	2·1 kg
Body material	Steel and concrete
Charge	TNT
Manufacturer	State factories

The Czech PP-Mi-Sb fragmentation mine consists of a short cylindrical casing made of concrete and small pieces of steel, normally erected on a wooden stake inserted into an aperture at one end, with the fuse projecting from the other. Usually an RO-1 pull fuse, attached to a tripwire, is used, but if required, the mine can be taken off its stake, placed in the ground, and an RO-2 pressure fuse substituted. In both cases a 1 kg force is needed to detonate the mine charge of 75 gm of TNT. If the tripwire is slack, the mine can be disarmed by severing the wire and extracting the fuse—a taut wire may be booby-trapped. The RO-2 fuse can be neutralised by inserting a pin in the safety-pin aperture and then extracting the fuse.

In service with the Czechoslovak forces.

PP-Mi-St anti-personnel mine

Czechoslovakia

Manufacturer	State factories

See the POMZ-2 and POMZ-2M.

In service with the Czechoslovak forces.

PT-Mi-Ba anti-tank mine

Czechoslovakia

Dimensions	Height 10·2 cm Diameter 32·3 cm
Weight	7·6 kg
Body material	Plastic
Charge	TNT
Manufacturer	State factories

The Czech PT-Mi-Ba anti-tank mine is of the plate-mine type, consisting of two round plastic mouldings welded together, with a smaller flattened dome in the centre on top (the pressure plate) and, on the base, a shallow fuse well in the centre, secured by a flat-threaded plate, flanked by two large flat bolt-heads to which a flexible plastic handle is attached. One of these boltheads is the filler plug, the other has no subsidiary purpose. A force of 200–400 kg applied to the pressure plate will cause the RO-7-11 fuse, located immediately below the plate and screwed into the rectangular booster charge, to shear and ignite the latter, which in turn detonates the main charge. This mine is not equipped with any integral booby trap and can, after clearing any external booby trap, be disarmed by unscrewing the threaded-plate from the base and removing the booster charge and shear-type fuse.

PT-Mi-Ba, underneath view

In service with the Czechoslovak and some Middle Eastern forces.

PT-Mi-D anti-tank mine

Czechoslovakia

Dimensions	Height 14 cm Length 32 cm Width 23 cm
Weight	9 kg
Body material	Wood
Charge	TNT
Manufacturer	State factories

The Czech PT-Mi-D anti-tank mine consists of a rectangular wooden box with a rope handle on each of two sides. In the centre of the top can be seen a rectangular aperture almost filled by the wooden pressure-plate, bevelled on all sides. The pressure-plate is supported inside the box by wooden dowels which shear when pressure is applied to the plate—this pressure can be adjusted from 200–450 kg, depending on the target in mind. When the dowels shear off, they release the shear blocks which are thus forced down on to the top of the striker, the retaining pin of the striker being knocked free, allowing the RO-1 fuse to ignite the 400 gm of TNT booster-charge which detonates the main charge of 6·2 kg of TNT. A popular method of booby-trapping the mine is to attach a wire to the retaining pin of the striker and secure the other end of the wire to the ground.

In service with the Czechoslovak forces.

PT–Mi–Ba III anti-tank mine

Czechoslovakia

Dimensions	Height 11 cm Diameter 33 cm
Weight	10 kg
Body material	Bakelite
Charge	7·5 kg
Actuation	Pressure release operating weight believed to be approximately 200 kg
Manufacturer	State factories

PT-Mi-Ba III anti-tank mine

The PT–Mi–Ba III bears the date of manufacture 1973. The mine is brown in colour, with a black rubber pressure ring and a light coloured plastic carrying handle underneath. Although the mine does not appear to have provisions for an anti-handling device, it could be booby-trapped by other means. This mine is undetectable by metallic mine detectors.

The striker mechanism is manufactured of brown plastic with the striker retainer, which is made of clear plastic, at the top of the mechanism. At the bottom of the striker mechanism is a threaded plastic portion which unscrews. The detonator is positioned through the hole till the lip of the detonator seats itself in the plastic portion, which is then screwed back into the striker mechanism.

The detonator is manufactured of aluminium with a purple cap. It is 45 mm long and 6 mm broad. The detonators are packed in wooden transit boxes each containing 15 detonators, with a dovetailed sliding lid.

The mine has a transit safety ring under the screw arming cap.

PT-Mi-K anti-tank mine

Czechoslovakia

Dimensions	Height 10·2 cm Diameter 30 cm
Weight	7·2 kg
Body material	Steel
Charge	TNT
Manufacturer	State factories

The PT-Mi-K is the most modern Czech mine and is equipped with booby-trap wells, which to a certain extent counteract the disadvantage of the metallic casing. Although approximately the same size and shape as the earlier plastic PT-Mi-Ba, the pressure-plate assembly is of a more angular profile and the steel carrying handle is hinged to the side of the casing. The PT-Mi-K consists of a circular steel casing filled with the main charge of 5 kg TNT, the centre being occupied by the pressure-plate supported by shear pins inserted in the rim. Directly underneath is the vertically-mounted fuse (either RO-5 or RO-9), also held by shear pins in a channel in the centre of the booster-charge. The booby-trap well is located centrally in the base of the outer casing. The mine is activated by 330–450 kg pressure applied to the pressure-plate, breaking the shear pins to allow the fuse to initiate the detonator situated below it. The mine may be disarmed by removing the pressure-plate and then the fuse, but the detonator may well be booby-trapped and should be left in place. It is believed that this mine was developed with nuclear warfare in view, and provision has been made to prevent over-pressure from a nuclear burst activating the pressure-plate.

In service with the Czechoslovak and some Middle Eastern forces.

K-2 anti-personnel bounding mine

Germany (DGR)

Dimensions	Diameter 10 cm
	Length 25 cm
Weight	5 kg
Body material	Plastic
Charge	Nitropenta
Manufacturer	State factories

Although the East Germans use Soviet mines, both anti-personnel and anti-tank, to a great extent, they have developed one of each of their own. Along with Soviet stake mines, they have produced the K-2 as their bounding mine. In appearance it resembles at first sight a stick-grenade with its short cylindrical body and long fuse housing. It is set by attaching a tripwire to the fuse, placing the mine base down on the ground and extracting the safety-pin from the fuse-housing. A pull on the tripwire causes the fuse to activate the propellant charge making the mine bound about 1·5 m in the air, where it discharges its contents of ball-bearings and steel pieces. Disarming is simply a matter of inserting a pin in the safety-pin aperture after first severing the tripwire (provided that the wire is slack). K-1, the anti-tank mine, was developed simultaneously in 1958.

In service with the DGR forces.

K-1 anti-tank mine

Germany (DGR)

Dimensions	Height 15 cm
	Diameter 25 cm
Weight	11 kg
Body material	Plastic
Charge	TNT
Manufacturer	State factories

Besides being constructed of plastic and thus difficult to detect, the East German K-1 anti-tank mine is equipped with a booby-trap fuse well. The K-1 is, with the K-2, a 1958 development to supplement the many Soviet mines used by the Nationale Volksarmee and has been retained in the inventory since then, with modifications only to fuse and detonator. Apart from the extra booby-trap fuse well, it is a conventional beehive-shaped mine, with the main pressure-plate on top in the centre and two small indentations on the side for the filler-plugs. In the base, the main fuse well is located centrally and the booby-trap fuse well to one side. This latter is a pull-fuse connected, when the mine is emplaced, to a short anchoring wire, and is very sensitive. It requires comparatively little pressure (150 kg) to detonate the charge of 7·5 kg of TNT.

In service with the DGR forces.

M49 anti-personnel mine

Hungary

Dimensions	Height 5·8 cm
	Length 18·5 cm
	Width 5 cm
Weight	0·3 kg
Body material	Wood or plastic
Charge	TNT
Manufacturer	State factories

The small Hungarian M49 anti-personnel mine was originally produced with a wooden casing, but is now more often encountered in plastic. This mine is a narrow box, resembling the Soviet anti-personnel PMD-7 mine, but longer and is made in two halves, the top fitting over the lower half, with the fuse projecting through an intersecting cut-away section at one end. The main charge of 75 gm of TNT is detonated by a pull-fuse and tripwire, taking a pull of only 1 kg.

In service with the Hungarian forces.

Bounding anti-personnel mine

Hungary

Dimensions	Height 30 cm
	Diameter 12 cm
Weight	3·6 kg
Body material	Steel
Charge	TRI-II
Manufacturer	State factories

After the immediate post-war period, during which she depended to a great extent on the Soviet Union for her mines, Hungary developed three mines, two anti-tank and this one anti-personnel bounding mine. The mine differs considerably in concept from other Warsaw Pact weapons of this type in that it is tripwire-operated whilst being mounted on a short stake. It consists of a cylinder topped by a short lid in the centre of which is set the rather long fuse attached to the tripwire. The cylinder is lined with ball-bearings packed round the explosive charge of 800 gm of TRI-II, which in turn surrounds the propellant charge of black powder. A pull of $4\frac{1}{2}$–9 kg (adjustable) on the tripwire will remove the safety pin from the fuse and release the two-pronged striker to activate the propellant charge which causes the mine to bound into the air and explode. Special provision is not made for booby-traps, but these may be present.

In service with the Hungarian forces.

Plate-charge anti-tank mine

Hungary

Dimensions	Height 8 cm
	Diameter 38 cm
Body material	Aluminium
Charge	PETN
Effect	Penetrates 100 mm of armour plate
Manufacturer	State factories

This is one of three mines developed by the Hungarians in recent years to supplement their holdings of Soviet mines. Like the other two, this weapon departs to a certain extent from the normal Warsaw Pact-type of mine design. The mine consists of a shallow aluminium bowl, three-quarters filled with PETN, capped with a broad aluminium ring, in the centre of which is bolted (with shear bolts) a circular steel disc 18 cm in diameter with a centrally-mounted steel spike. The electrically-operated fuse is located in the base of the bowl. Normal procedure is not to lay the mine in the ground, as with most other anti-tank mines, but to prop the mine up in a wooden triangular framework facing the direction of enemy advance. When a target reaches a range of about 20 m the mine is detonated by remote control—the projectile, consisting of the disc and spike, is said to be capable of piercing 100 mm of armour at that range. This mine would appear to be limited in scope except where traffic approaches are severely restricted and undergrowth (for camouflage) is plentiful.

In service with the Hungarian forces.

Shaped-charge anti-tank mine

Hungary

Dimensions	Height 14·2 cm
	Diameter 30 cm
Weight	5·7 kg
Body material	Cardboard, wood and canvas
Charge	TNT
Manufacturer	State factories

This, the third of the Hungarian-developed post-war mines, must be one of the flimsiest anti-tank mines made in recent years; one of the cheapest to produce, but nonetheless effective. It consists of a round box, the side of cardboard, the top and bottom of thin wood and the whole encased in a camouflage-painted canvas cover. Set erect in the centre of the base is a pressure fuse, projecting into the TNT shaped charge which is encased in oiled paper and suspended from the underside of the circumference of the top. Although the fuse is normally set to be activated at about 450 kg (adjustable), it is surmised that it would need considerably less effort to crush this device and damage the fuse. Perhaps a more effective method would be to use the system, which apparently is practised, of attaching the mine to a loose projecting angle-iron. When a vehicle track or wheel strikes the angle-iron, the mine is thrown up violently against the vehicle body and detonates.

In service with the Hungarian forces.

PMK-40 anti-personnel mine

USSR

Dimensions	Height 3·8 cm
	Diameter 7 cm
Weight	90 gm
Body material	Cardboard or thin metal
Charge	50 gm TNT
Manufacturer	State factories

This tiny nuisance-value mine contains only 50 gm of TNT and is intended to be left strewn around in abandoned locations to be stepped on. It consists of a little, round, waxed cardboard or thin metal box with a bevelled top and a small round filler-plug on the side. In the box are the TNT and a pressure-operated lever fuse which, when subjected to about 9 kg of pressure, frees the spring-loaded striker to ignite a percussion-cap. The PMK-40 is hardly likely to kill, but can cause a nasty foot injury, and it is not easily noticed in grass or undergrowth. Furthermore, only the fuse assembly is of metal and the mine could easily elude a conventional metallic mine detector.

In service with Soviet and the Vietnamese forces.

PMN anti-personnel mine

USSR and China

Dimensions	Height 5·6 cm
	Diameter 11·2 cm
Weight	0·6 kg
Body material	Plastic
Charge	216 gm TNT
Manufacturer	State factories

PMN anti-personnel mine

The Soviet PMN anti-personnel mine is a little larger than the PMK-40 mini-mine, but performs the same function, using similar operating procedures. Indeed, it was developed as the successor to the PMK-40. The PMN consists of 240 gm of TNT, percussion-cap detonator and spring-loaded striker, encased in a small, round, flat duroplastic box. When the safety pin is withdrawn the igniter moves slowly into the mine until it is stopped by a step in the centre. The top and bottom halves of the case are held together by a rubber pad fastened by a small wire clip, and on the side of the bottom half there is a comparatively large cap, protecting the fuse and detonator, projecting through an aperture. As with the PMK-40, this mine is intended to be left strewn around in areas abandoned to the enemy, detonation occurring (with rather more severe results than with the PMK-40) when a soldier (or civilian) inadvertently treads on one. The resultant pressure forces the spring striker to ignite the percussion-cap. In the base of the PMN are four shallow wells formed by cruciform ribs, providing the opportunity for further booby-trapping.

In service with the Soviet and Chinese forces.

OZM series anti-personnel mines

USSR

Main constituent	120 mm mortar bomb or 122 mm/152 mm artillery shell
Charge	TNT
Manufacturer	State factories

The OZM anti-personnel mine consists of a certain type of projectile, modified by the attachment of, in place of the nose-mounted impact fuse, a small box-shaped propellant charge (the UVK-1). This is connected by a thin metal tube to the fuse mounted at right-angles to the tail fins (or shell base). The fuse can be of either the pressure, pull or electrical type, and performs two tasks—activation of the propellant charge and of the delayed-action system of the detonator. The OZM is usually buried in the earth, nose down, with only the fuse lying above the mine and parallel to the ground. When the fuse is activated, the propellant projects the mine into the air between $1\frac{1}{2}$–$2\frac{1}{2}$ m, depending on the size of the bomb or shell used, and simultaneously ignites the delayed-action system incorporated in the propellant charge-box to detonate the mine at the top of its jump. The amount of explosive charge contained varies with the type of projectile, the 120 mm mortar bomb carrying 1·8 kg of TNT, the 122 mm shell 3·4 kg and the 152 mm shell nearly 6·3 kg, while the complete mines weigh 18·6 kg, 24·5 kg and 45·4 kg respectively. It is therefore clear that such weapons can be a hazard to armoured vehicles as well as to infantry.

In service with the Soviet and Vietnamese forces.

OZM3 anti-personnel mine

USSR

Weight	4·5 gm
Body material	Metal
Charge	0·75 gm
Manufacturer	State factories

This jumping mine is unusual in that the hole in the ground in which it is placed acts as its outer pot.

The OZM3 consists of a steel cylinder with a thin tinplate base with a centrally placed key for a detonator well. On one side of the key is a raised threaded adaptor and on the other side are two electrical leads.

Based on the UVK propellant system, it may be detonated by a trip-wire operated igniter or by command detonation. A flash caused by one of the ignition systems fires the propellant charge. This detonates a delayed fuse which ignites the main charge. The type of soil and the type of hole dug will govern the height the mine is thrown into the air. The charge throws metal, from an inner fragmentation shell, up to a radius of 45 m.

In service with the Soviet forces.

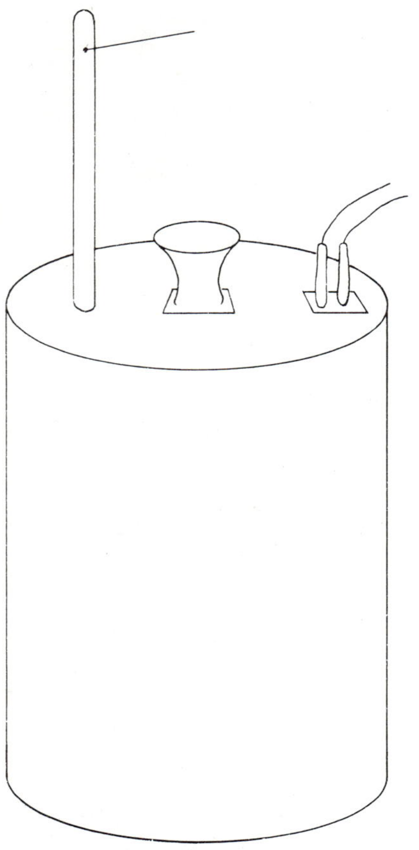

OZM-3 anti-personnel mine

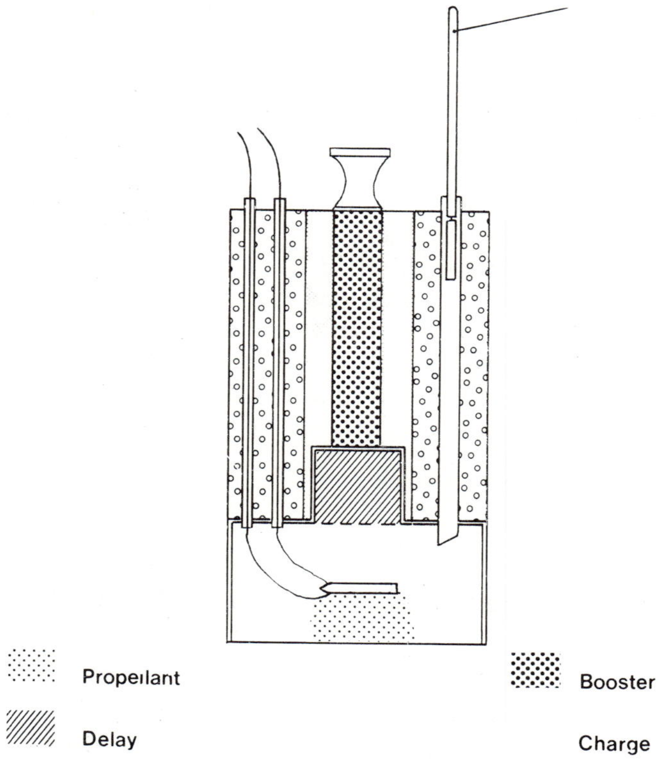

PMD-6 anti-personnel mine

USSR

Dimensions	Height 6·5 cm Length 20 cm Width 7·5 cm
Weight	0·4 kg
Body material	Wood
Charge	TNT
Manufacturer	State factories

The Soviet PMD-6, of similar design to the Czech plastic PT-Mi-D, is an anti-personnel blast mine consisting of a wooden two-piece casing, the top folding on to the bottom, containing a 200 gm block of TNT and a pull-type MUV fuse which projects through intersecting apertures cut in one end of both casing sections. Although it is possible to use this mine with a tripwire, it is more effective to set the lid so that a soldier can set off the mine by stepping on it (or placing something on top of it). In this way, the striker-retaining pin is pulled out by the downward movement of the edge of the cover. In a similar manner, it is possible to booby-trap the PMD-6 by attaching a cord or wire to the striker-retaining pin and anchoring the other end of the string to the ground. This mine is of such simple design that it is perfectly feasible to make one in the field and, consequently, although the explosive power will remain about the same, the dimensions may vary slightly. (See also PMD-7, PMD-7ts and PT-Mi-D.) Pressure required to detonate these mines varies from 1 kg to 9 kg.

In service with the Soviet, Vietnamese and North Korean forces.

PMD-7, PMD-7ts anti-personnel mines

USSR

Dimensions	Height 5·1 cm Length 15·2 cm Width 7·6 cm
Weight	0·3 kg
Body material	Wood
Charge	TNT
Manufacturer	State factories

The PMD-7 and PMD-7ts Soviet blast anti-personnel mines are identical except that the lower casing of the latter model is cut out of a single wooden block. The working principles and method of employment are the same as those of the PMD-6, which is larger and heavier and contains more TNT (200 gm as against only 75 gm). Again, the amount of pressure needed to detonate the mine can vary from 1 kg to 9 kg.

In service with the Soviet forces.

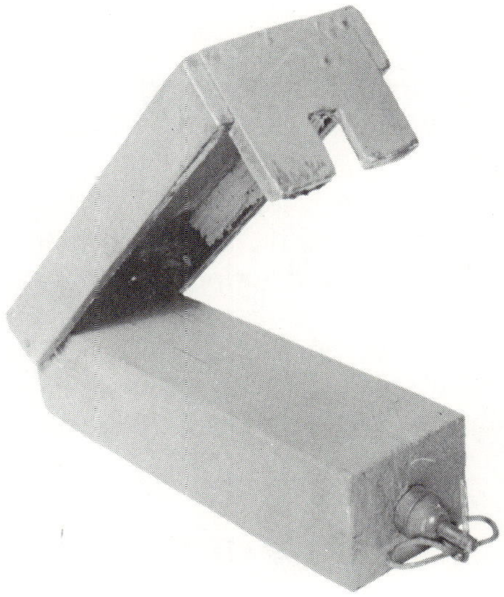

PMD-7ts anti-personnel mine

POMZ-2 and POMZ-2M anti-personnel mines

USSR

Dimensions	Diameter 6·4 cm
	Length 12·7 cm
Weight	2 kg
Body material	Cast iron
Charge	7 gm TNT
Manufacturer	State factories

The POMZ-2 pre-fragmented anti-personnel mines rather resemble a Mills hand grenade planted on top of a wooden stake. Both models consist of a short, cylindrical, cast-iron segmented body with the pull-type MUV or MUV-2 fuse projecting from the top, and a charge of 75 gm of TNT. The usual system, as was practised for years along the borders of Iron Curtain countries with Western countries, is to attach tripwires to the mines and to set them out in groups. The older POMZ-2 is heavier and longer in body than the modernised version, but contains the same amount of explosive and, in both cases, a pull of 1 kg is sufficient to cause detonation, severely injuring any person within 20 m range. Besides the Chinese, the Czechs and Yugoslavs also make this mine, calling it the PP-Mi-St and PP-MR-S respectively.

In service with the Chinese, Soviet, Vietnamese, most Warsaw Pact and Yugoslav forces.

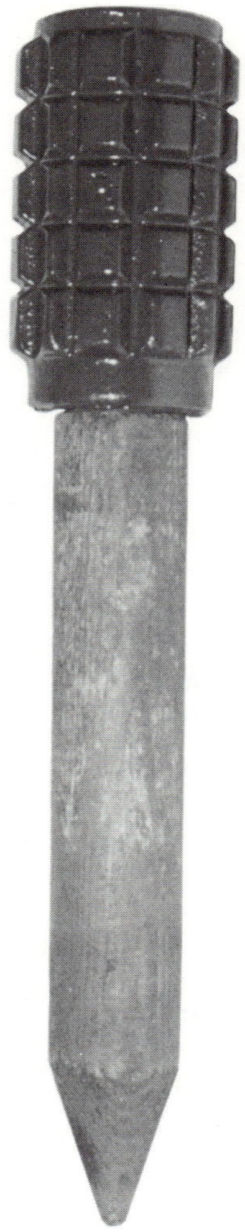

POMZ-2 anti-personnel mine

LMG anti-tank (grenade) mine

USSR

Weight	9·5 kg
Body material	Steel
Charge	TNT
Manufacturer	State factories

The literal translation of the title LMG—'the Galitskii Flying Mine'—is a good description of its function. The LMG is really a tripwire-operated rocket, fired from a simple wooden anchored frame. The mine consists of a sheet-steel cylindrical body with a conical nose and base, which contains the 187 gm shaped charge of TNT, and a thin cylindrical tail section with four short delta-shaped fins, containing the MUV pull-type fuse and the black-powder propellant. Firing as it does only on a fixed line, this mine would appear to be only really effective if used in locations where traffic approaches are very restricted and sufficient camouflage is available. However, these mines were used quite effectively by the Red Army in World War II.

Probably held in reserve by the Soviet forces.

TMB-2 anti-tank mine

USSR

Dimensions	Height 13·9 cm
	Diameter 26·7 cm
Weight	6·6 kg
Body material	Tarred cardboard
Charge	5 kg Amatol (or TNT)
Manufacturer	State factories

The TMB-2 (and also the similar, but slightly larger, TMSB) anti-tank mine resembles a large chocolate cake, being a round, dark-brown box with a black tape round the middle sealing the top and bottom together and a small, blue glass disc (the screw-plug of the fuse well) on top in the centre. Two-thirds of the mine contain the main charge, of about 6 kg HE, in the centre of which is placed vertically a 75 MD-2 detonating cartridge, and in the top part of that is inserted the MV-5 pressure-fuse, set to operate under a force of only 11·8 kg. Efforts have been made to combine the mine's non-detectability (through avoidance of metal parts) with resistance to detonation (through the use of tar, asphalt and tape) to such effect that this mine has for a while been accepted as standard issue in the Soviet Army, in spite of its apparently flimsy structure. The main charge can vary from 5 kg–6·5 kg and may be either TNT or amatol.

In service with the Soviet and Vietnamese forces.

TMSB anti-tank mine

USSR

Dimensions	Height 16·5 cm Diameter 28·6 cm
Weight	7 kg
Body material	Tarred cardboard
Charge	6 kg Amatol
Manufacturer	State factories

The TMSB anti-tank mine is similar in appearance to the TMB-2 and is identical in its construction and function. It is, however, slightly larger, and the main charge is 6 kg of amatol. Otherwise, characteristics are the same.

In service with the Soviet forces.

TMSB anti-tank mine

TMD-B and TMD-44 anti-tank mines

USSR

Dimensions	Height 14 cm Length 32 cm Width 28 cm
Weight	7·7 kg (average)
Body material	Wood
Charge	Amatol, dynammon, ammonite, TNT, picric acid, or others
Manufacturer	State factories

The well-tried TMD-B wooden anti-tank mine (very similar to the Czech PT-Mi-D and the Soviet TMD-44) is housed in an almost square box with a canvas handle on one side and a lid consisting of three short planks set side by side in the centre. These planks form the pressure-plate and the centre plank is hinged to allow the MV-5 pressure-fuse to be inserted in the MD-2 detonator. When this has been done, the centre plank is locked in place by a narrow slat pushed through a groove in the ends of all three planks. Inside the box the main charge can consist of 5 kg to 7 kg of any of the explosives mentioned above, but the booster-charge is usually a 200 gm block of TNT. All specifications of the TMD-B can vary,

TMD-B anti-tank mine

as the simple design allows this mine to be constructed in the field as well as industrially, but the force required to detonate the mine is usually about 200 kg. The TMD-B was employed by the Red Army in the latter stages of World War II and was one of several used by the North Koreans against United Nations troops, prior to its appearance in Indo-China.

In service with the Soviet, North Korean and Vietnamese forces.

TMD-44 is an improvement on the TMD-B wooden mine, but adopts similar operating principles. It can be booby-trapped and converted to an AP mine. The picture shows a Cuban version with a bakelite fuse well-cover, an MV5 switch and an MD2 detonator

TM46, TMN46 and TM57 anti-tank mines

USSR and China

Dimensions	Height 7·4 cm
	Diameter 31 cm
Weight	8·7 kg
Body material	Sheet steel
Charge	5·7 kg TNT
Manufacturer	State factories

The only difference between the Soviet TM46 and TMN46 anti-tank mine is that the latter is equipped with a booby-trap fuse well set to one side in the base. Both mines should always be treated when found, therefore, as being of the TMN46 type and fitted with an anti-lift system. These mines are round, flattish, steel casings, painted green, with a slightly domed, ridged top, containing, in the centre, a scalloped cap for the MV-5 pressure-fuse well. The filling-plug is located on the side beneath the light metal handle. The mine contains 5·7 kg of TNT, detonated by a centrally-placed 200 gm block booster-charge, also of TNT, which in turn is activated by a MD-6 detonator attached to the MV-5 igniter,

The TM46 anti-tank mine

TM46 anti-tank mines, Chinese version

is inserted in the booster-charge and connected directly to the fuse—a 180 kg force is sufficient to set off the mine. Although these mines were designed for use with the PMR-3 and PMR-60 mine-laying chutes, it is quite feasible to lay them by hand, either in groups or as a single weapon.

In service with the Chinese, Soviet and Vietnamese forces.

The Soviet TM57 mine (left) with the MB3-57 fuse removed and the TM46 mine (right) with the MVM fuse removed. Both mines show a state of diaphragm collapse caused by traffic.
Many experts believe that the MVM fuse is built as a single unit for safety to prevent the uninitiated from inserting the igniter mechanism upside down, but this is not entirely correct. It is a mechanical laying safety device to decrease risk of premature detonation whilst the laying party is in the immediate area

The MB3-57 mechanical self-arming fuse for the Soviet TM57 mine. The left photograph shows the fuse with the safety ring still in position and the photograph above shows the clockwork mechanism (in safe state). The wire safety ring is lifted up and pulled back, removing the half moon safety clip from under the activating button. When the button is depressed, the clockwork motor inside the cap is activated causing the detonator unit to turn 90° from the safe horizontal position to the vertical position, bringing the detonator into line with the booster inside the mine itself, thus completing the arming chain. Depending on how much the clockwork motor is wound, it takes between 20 and 80 seconds to arm. It is assessed that this safety device has been designed to decrease the risk of premature detonation when mechanically laid. When the detonator unit is transported, a brown plastic cap is screwed on to avoid damage to the mechanism

YaM-5 anti-tank mine

USSR

Dimensions	Height 15·2 cm Length 50·8 cm Width 17·9 cm
Weight	5·9 kg–7·7 kg
Body material	Wood
Charge	3·5 kg–5 kg Amatol (or TNT)
Manufacturer	State factories

The Soviet YaM-5 anti-tank mine, like most wooden mines, is manufactured in several slightly different sizes but, unlike many of the others, is not sufficiently simple to be easily constructed in the field. Additionally, there is a rather larger version, the YaM-10. The case of this mine is a rectangular box with a hinged lid, which is provided with a flap-extension down the front of the box. The open-ended aperture in the centre of this flap serves to admit the end of the MUV pressure-fuse striker, which is held in position by a transverse pin between two staples affixed to the flap. A force of 136 kg on the box lid allows the staples to release the retaining pin and activate the fuse to detonate the main charge of 3·5 kg to 5 kg of HE (depending on the type of explosive and size of mine). There are no booby-trap fuse wells built into the YaM-5, but it is possible that an anti-lift anchoring wire could be attached to the striker retaining-pin.

In service with the Soviet forces.

YaM-10 anti-tank mine

USSR

Dimensions	Height 20 cm Length 62 cm Width 22 cm
Weight	11·8 kg
Body material	Wood
Charge	7·2 kg–10 kg Amatol (or TNT)
Manufacturer	State factories

The design and operating principles of the YaM-10 are the same as for the YaM-5, which it is replacing. The only differences between the two are a matter of size, the YaM-10 being virtually half as big again and containing a greatly increased main charge of 10 kg of HE.

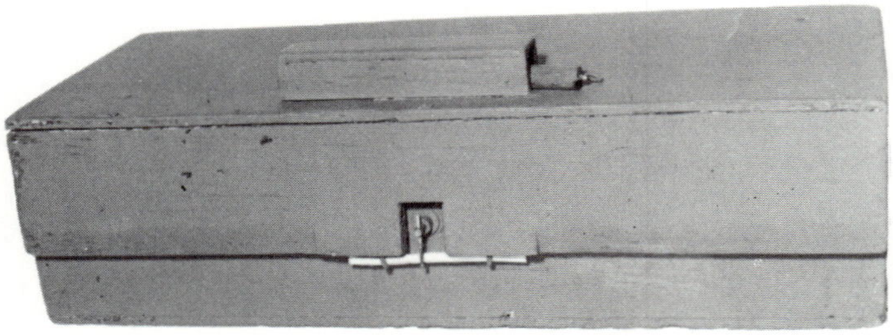

YaM-10 anti-tank mine

In service with the Soviet forces.

Dog anti-tank mine

USSR

Weight	6·5 kg
Body material	Canvas
Charge	TNT
Manufacturer	State factories

The Red Army learned to use these mines from the Japanese and the German armies in World War II. The system comprises two 3 kg TNT blocks housed in the two pouches of a canvas jacket, 200 kg of booster-charge, and a small metal box, located in the centre, containing the MUV pull-type fuse and operated by a trip-lever about 20 cm in length. The systems are mounted on the backs of dogs specially trained to run below the tank belly between the tracks when released by handlers. As a dog runs under the tank the trip-lever catches on the under side of the tank and releases the retaining-pin of the fuse to detonate booster and main charges. Even today there is still said to be a shortage of dogs in the USSR.

In service with the Soviet forces.

DM general purpose mine

USSR

Dimensions	Height 13 cm Length 15·5 cm Width 15·5 cm
Weight	1·8 kg
Body material	Wood
Charge	1·2 kg TNT
Manufacturer	State factories

The DM is a small vibration mine normally used buried in a loose sand surface. Whilst not big enough in itself to be a danger to tanks, it can be used in the anti-tank role to trigger off a larger charge. In appearance, it is a square grey box of plywood with a sliding lid and two small holes in one side where wires project to connect with the battery of the electric detonator. The inside of the box is divided into halves, the bottom half containing the main charge of 1·2 kg of TNT, and the top half housing the large, circular, clockwork ChVZ vibration fuse and electric blasting-cap. This mine is a booby-trap in itself as, once armed, it cannot be neutralised and must be exploded from a safe distance. Direct pressure is not necessary to detonate the mine, which can be set off merely by the close passage of a motor vehicle or even an animal-drawn wagon.

In service with the Soviet and Vietnamese forces.

MZD general purpose mine

USSR

Dimensions	Various (rectangular)
Weight	Various
Body material	Wood
Charge	Amatol, TNT, ammonite, dynammon, picric acid, or others
Manufacturer	State factories

The Soviet MZD mines are housed in plain rectangular wooden boxes with sliding tops. They vary in size, depending on the amount of explosive they contain—usually between 400 gm and 10 kg. Their purpose is equally varied, including anti-tank, anti-soft-skinned vehicles, for blowing up trains, or even as booby-traps in vacated buildings. Consequently, the fuses used will also vary in accordance with the purpose—VZ-1 (vibration), ChZ-35 or CRZ-10 (electric delay), EKRP (safety delay) or EKRV (electric) fuses can all be used. As the vibration and electric fuses are extremely sensitive, no attempt should be made to neutralise these mines by hand—the type of fuse cannot be discerned by visual inspection.

In service with the Soviet forces.

AKS general purpose mine

USSR

Dimensions	Height 12 cm Length 27 cm Width 23 cm
Weight	9 kg
Body material	Sheet steel
Charge	TNT
Manufacturer	State factories

The Soviet AKS tilt-rod-operated mine has been proved to be more reliable than many other newer mines, and is effective against both men and vehicles. The main charge of 6·8 kg of TNT is contained in a square steel box, one side of which is a sliding panel, with, mounted centrally on top in a ball-joint, the tilt-rod. The latter is connected inside the case to the retaining-pin of the MUV pull-fuse, but is secured against inadvertent movement by a safety pin. Although the 63 cm long tilt-rod can be left in well-camouflaged cover to catch the underside of a vehicle on its own, a common practice is to bury the mine-casing and attach a large twig to disguise the rod. The amount of pressure and tilt required to detonate the mine can vary from 5 kg to 20 kg and 20°–50° respectively. Booby-trap wells are not provided, but an anti-lift device may be improvised.

In service with the Soviet and Vietnamese forces.

Magnetic (non-metallic) limpet mine

USSR

Dimensions	Length 26·7 cm
	Diameter 11.4 cm
Weight	2·8 kg (without fuse)
Fuse	Mechanical delay
Main charge	TNT with high aluminium content and RDX booster
Manufacturer	State factories

This is a moulded, plastic, semi-cylindrical limpet mine which uses a mechanical time delay and which can be used for disabling ships or can be attached to fuel drums, armoured vehicles or any other suitable target. The fuse delay can be set from 5 minutes to 823 hours by using one of the six delay tabs. The initiating device is an aluminium mechanical time device operated by spring pressure—a pull of the safety ring releases a strong spring, but the force of the spring is restrained by a wire which passes around the delay tab. Depending on the temperature and the thickness of the tab, the wire takes a predetermined time to cut through tab. After the wire has cut through the tab, the spring is released and drives the firing pin into the detonator, exploding the mine.

The mine has a horseshoe magnet at each end and a double fuse well at one end with a black plastic screw-threaded plug. The magnets provide a stand-off of 16 mm. The mine has grey coloured transit plate and green fabric carrying handle.

The fuse is packed in a round cardboard container. Enclosed in the container are lead delay strips with the time chart, ie colour code with fuse, also enclosed is a locking key or spanner.

Mine and fuse

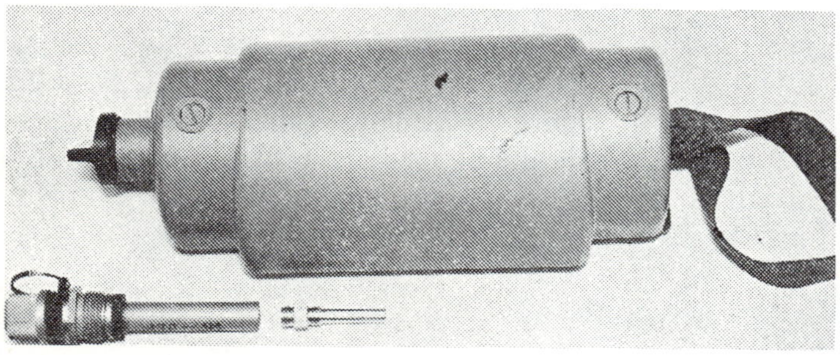

Underside of mine with transit plug and transit plate in position

PMA3 anti-personnel mine

Yugoslavia

Dimensions	Height 3·7 cm
	Diameter 10 cm
Weight	185 gm
Body material	Plastic
Charge	15 gm Amatol

The new Yugoslav anti-personnel mine—the PMA3—weighs only 185 gm and contains a minute 15 gm of Amatol, which is insufficient to cause serious injuries. It is made of two dull apple-green plastic components and when assembled is held together by a black rubber cap. A green plastic spring-loaded safety band is situated around the mine, exerting equal pressure around the circumference and locking the two parts together to prevent movement.

To arm the mine, a chemically activated detonator is inserted into the well in the bottom section by removing the plug. Replace the plug and gently remove the spring-loaded clip and safety band by pulling the 1 m nylon cord. The spring and cord will then force the band out from between the two components, thus allowing the components to move. It is initiated by pressure (probably as little as 2 kg) being applied onto the rim, causing the two components to pivot on each other, thereby shearing the chemical detonator. It is assessed as a nuisance mine, to be laid at random in abandoned areas or to create cordons-sanitaires.

PMA3 anti-personnel mine dismantled

PMA3 (right) and PMN anti-personnel mines

TMA3 anti-tank mine

Yugoslavia

Dimensions	Height 8·2 cm Diameter 26·5 cm
Weight	6·6 kg
Manufacturer	State factories

The TMA3 is entirely non-metallic in construction and cannot therefore be detected with the usual metallic mine detectors.

One side has three detonator wells set near the edges of the circumference of the mine. These are approximately 1·9 cm in diameter and three plastic detonator assemblies are screwed into them. A single detonator well is centrally situated on the opposite face and is generally believed to have an anti-handling device. It also has a webbing loop attached at one side for carrying purposes.

Pressure required to initiate igniters is not known at this stage, but is believed to be in the region of 100 kg.

TMA3 anti-tank mine. View showing equidistant exposed fuse wells

Three TMA3 detonators in carrying-case

TMA4 anti-tank mine

Yugoslavia

The current Yugoslav non-metallic vehicle mine—the TMA4—is an improvement on the TMA3. Whereas the TMA3 has a brown varnished hessian skin round the TNT charge, the TMA4 has a green plastic cover and the mine designation painted in yellow on one side. The TMA4 uses a slightly modified igniter mechanism, but incorporating the same ignition compound as the TMA3. It has three detonator wells on one side and one on the reverse. It has a canvas carrying handle moulded into the mine. Weighing only 6·3 kg, it requires about 10 kg to initiate the detonator. Due to its complete absence of metallic parts, it is impossible to detect with conventional metallic mine detectors.
It is believed that the Yugoslavs now have an improved version of the TMA3 and TMA4 on their inventory of anti-tank mines. No information is as yet available.

TMA4 anti-tank mine

Cake tin mine

A new, and as yet unidentified, anti-tank mine exists which is probably of Soviet origin and which looks exactly like a cake tin—hence its nickname. The container is made of a light coffee-coloured PVC/plastic.

The explosive (TNT) is cast up to the top of the sides of the mine which has three cavities to house three AP mines. The lid then fits firmly on to the top sitting flush with the AP mines. This concept of using AP mines as igniters is not new in the practice of mine warfare, but it is unique in that it appears to be the first state manufactured mine to incorporate the idea.

The cavities differ slightly in size to take the different AP mines available. Mines have been found with Soviet PMN AP mines and United States M14 AP mines.

Combat Aids

This section deals with numerous miscellaneous pieces of equipment, some of them technically far in advance of the weapons in the earlier part of the book. Many of them are very similar and, for this reason, a relatively small selection of each has been chosen, more to illustrate their existence than to catalogue them for the researcher. Indeed it is difficult to know where to draw the line. In this edition we have drawn it above anti-terrorist devices. Even so, a number of combat aids may well have been omitted.

Considerable ingenuity is shown in the development and manufacture of combat aids, and it is interesting to note how some old ideas are re-thought and reproduced in a new form. Others are straightforward inventions or developments which, for example, have done much to extend the soldier's detection range. The situation in this sphere is perhaps becoming similar for the infantry as it has been for the cavalry in another. It would seem that there is considerable similarity between the escalatory development of the tank and the anti-tank gun with that of the range of small arms and the firer's ability to see and detect further, especially at night.

Bayonets

For many years, the practical use of the bayonet has been under review, but, in the meantime, manufacturers continue to offer this accessory almost without exception.

The bayonet is as old as the gun and, as such, has gone through almost as many changes within its limitations. The original concept of the bayonet was purely practical, namely, to have a secondary weapon available when the gun had been discharged. However, from the early 1900s, bolt actions with magazines which could be rapidly reloaded became available. This factor rather reduced the need for the bayonet. Nevertheless, it was retained.

During World War I, it was felt that the psychological effect on the enemy when faced with the bayonet was so great that at one time General Haig ordered a major offensive to commence at sunrise, rather than at first light, in order that the sunlight should glint off the polished steel of the men's SMLE bayonets. In practice, in the confined spaces of the trenches, the long-bladed bayonets were too clumsy and ineffectual, and a sharpened spade (or similar implement) was considered much more effective.

In the period between the wars, some countries began adopting shorter rifles and correspondingly shorter bayonets. Some rifles were even fitted with folding bayonets, though this again was not a new idea. With the shortening of the bayonet, the weapon began to come into its own once more as it could be used as a fighting knife. However, the British No 4 Mk 1 rifle bayonet was a socket type and with a cruciform cross-sectional blade. Totally useless as a knife, a version was produced with a swivelling pommel and Bowie-type blade to fill the obvious gap that was felt by the infantryman. It is worthy of note that the Germans had by this time removed the traditional quillon from their 98k bayonet, feeling it to be an unnecessary expense.

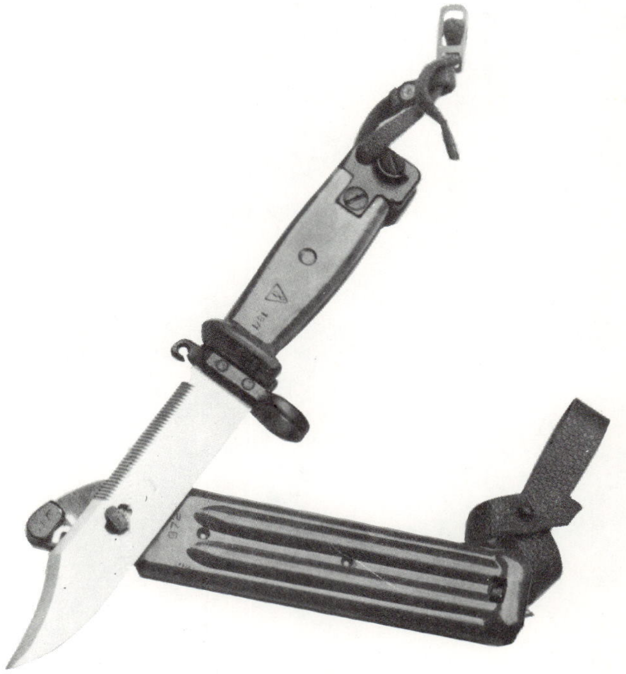

Russian bayonet for the Kalashnikov AKM rifle

Different types of blade were fitted for different uses, the most common innovation being the saw-back blade for pioneer use. Different accessories have also been attached to the bayonet such as grenade discharger cups or devices for cutting barbed wire by feeding the wire along the bayonet to the muzzle and thus firing the rifle to cut the wire, or bottle openers.

Nowadays, the bayonet is being designed more as a general purpose tool rather than merely a weapon. It also has its place on ceremonial parades!

Some of the more modern bayonets such as those of the Russian AKM and the French M49/56 are made so that the blade may be hinged on a stud attached to the scabbard and, using a corresponding cutting edge on the outside of the scabbard, the bayonet may then be used as a wire cutter. These bayonets are also capable of floating and feature a thong enabling them to be secured to the leg. Other bayonets produced recently, incorporate screw-driver points and containers. Thus it would appear that the wheel has turned a full circle, and the bayonet has become a useful weapon and combat aid.

Mine Detectors

In simple terms, there are four ways of detecting mines—by electronic equipment, by prodding, by using mine dogs and, in the apocryphal manner by sending one's wife on ahead. This section deals with the first two, which are the most successful and widely used.

VISF M1946 mine detector

Bulgaria

Mounting	Manpack
Weight	6·2 kg
Manufacturer	State factories

During World War II, the Bulgarians used both Russian and German mine detectors and, using these as a basis, developed their own, of which the man-portable VISF M1946 is the most likely to be still in use. The VSIF M1946 was developed from the Soviet VIM203, which is now obsolete, but is only just over half as heavy. It consists of a hollow, wooden, square detector-head frame with a metal covering, and a hollow wooden handle attached to the head by an adjustable hinge—at the other end of the handle and attached parallel to it is a large, rectangular battery box containing the power supply of a $1\frac{1}{2}$ volt dry battery and a 60 volt wet cell battery. The headphones are connected by a short cable to the tone-regulator halfway along the handle and from there down the inside of the handle to the aluminium amplifier box, which contains a single pentode amplifier and twin pentode oscillators. In use, the VISF M1946 detector's handle can be extended by an extra three sections and for travel is carried in a canvas pack. Operation is on the beat-frequency oscillation principle, and it is effective only against metallic mines.

In service with the Bulgarian forces.

M10 and M11 mine detectors

Czechoslovakia

Mounting	Manpack
Weight	12 kg
Manufacturer	State factories

The Czech M10 and M11 portable mine detectors are two very similar devices operating on the conventional beat-frequency oscillation principle. In both cases the tuning-box is carried in a canvas pack on the back of the operator, whose headphones are connected to both tuning-box and the four-piece aluminium extending handle by wires which pass down the outside of the handle (not inside) to the detector-head. The construction of the head forms the sole difference between the M10 and M11 models: that of the M10 consists of twin overlapping discs, with the handle jointed to the centre of the top disc—the M11 detector-head is a solid figure-8 shape of the same size and with the handle jointed in the same relative position. The tuning-box in the pack also contains the tone-regulator and the three 25 volt dry batteries which, when not in use, are kept in a small pouch on the side of the pack. Only metallic mines can be discovered by these two detectors.

In service with the Czechoslovak forces.

MSG64 mine detector

Germany (DGR)

Mounting	Manpack
Weight	4·4 kg
Manufacturer	State factories

This modern lightweight mine detector is a typical instance of effective German electronic engineering. Whilst the detector itself weighs less than 2·5 kg, it is capable of locating tiny metal objects, even below water up to 18 cm away from the detector-head. As with other conventional mine detectors, the MSG (Minensuchgerät) 64 operates on the beat-frequency oscillation principle. This detector consists of an oval coil detector-head with a four-section aluminium handle which can be extended to a length of 2·4 m. All controls and accessories are directly attached to the handle section permanently fixed to the detector-head—the four $1\frac{1}{2}$ volt dry batteries for the power, the tone-regulator and the timing-box with its twin oscillators. The headphones are connected by wire to the tone-regulator. For carriage when not in use the MSG64 is taken apart and stored in a canvas pack.

In service with the DGR forces.

UMIV-1 mine detector

USSR

Mounting	Manpack
Weight	6·6 kg
Manufacturer	State factories

The Soviet UMIV-1 detector resembles the East German MSG64, but is considerably heavier, has a much greater range and does not function under water. Like the MSG64, it has an oval-coil detector-head and the tuning-box with its twin oscillators and the tone control are built into the part of the four-section collapsible handle permanently attached to the detector-head. The operator, however, has additionally to carry a large rectangular pack, holding the 3 volt filament and 70 volt plate power supply, to which both his headphones and the free end of the detonator handle are connected by cable. If the handle is fully extended it will reach 1·3 m, and the detector has a range of 45 cm. The UMIV-1 operates on the beat-frequency oscillation principle and is only capable of discovering metallic mines. It is being replaced by the more modern IMP detector.

In limited service with the Soviet and some Warsaw Pact forces.

UMIV-1 mine detector in action

IMP mine detector

USSR

Mounting	Manpack
Weight	9·7 kg
Manufacturer	State factories

The replacement for the UMIV-1 portable mine detector is the modern transistorised IMP manpack detector, which not only has the capability of discovering any metal object down to a depth of 1 m, but can operate with equal facility under water. In appearance it is totally different from other Warsaw Pact detectors, the detector-head being housed in a plastic cylinder 42 cm × 4 cm, set at right angles to the collapsible four-part aluminium handle, which is almost 2 m long when fully assembled. The controls and batteries of the IMP are not located on the handle, but in a small rectangular box, comprising the tone-regulator set over the tuning control, working on five transistors powered by four $1\frac{1}{2}$ volt dry batteries—2·5 kg in weight. The headphones and the detector-head are connected directly by cable to the control box. For travel, and when not in use, the whole assembly stows away in a light rectangular metal box.

In service with the Soviet and some Warsaw Pact forces.

IMP mine detector in action

DIM mine detector

USSR

DIM vehicle-mounted mine detector

Mounting GAZ/UAZ 69 truck
Manufacturer State factories

The DIM mine detector, mounted on the Soviet GAZ/UAZ 69 jeep, was designed originally to clear roads but can, of course, also be used to clear a projected camp site. Once again, however, only metallic mines can be detected by this equipment. The detector-head is formed by the thick tubular front portion of the two-wheeled frame carried on the bonnet of the vehicle. To operate, the frame, hinged to the bumper bar, is lowered to the front on to its own two wheels with the detector-head in front of the wheels a few cms from the ground. If the head discovers any metal object in the ground down to a depth of about 30 cm, an automatic electrical impulse applies the vehicle brakes. But as the detector-head is large it is unable to pinpoint the position of the object—a normal portable detector is therefore an essential accessory. For travel, the whole frame is raised and carried over the vehicle bonnet. A working speed of around 5 kph is possible.

In service with the Soviet and other Warsaw Pact forces.

Mine probes

USSR

Manufacturer State factories

Conventional mine detectors usually have to function on electric power from batteries. It therefore follows that when batteries run flat or are not obtainable, as can easily happen under operational conditions, some other method must be found to search for mines. The Red Army during World War II developed a selection of probes of varying shapes and sizes for this purpose, simply to probe the earth to locate mines. The following examples were then standard and it can be supposed that, if necessary, would be so again:

(1) a simple metal probe like a long fire poker;

(2) a simple wooden pole with a metal rod projecting from the end;

(3) a multi-section wood or metal pole with a projecting rod;

(4) a multi-section probe which folds up rather than takes apart;

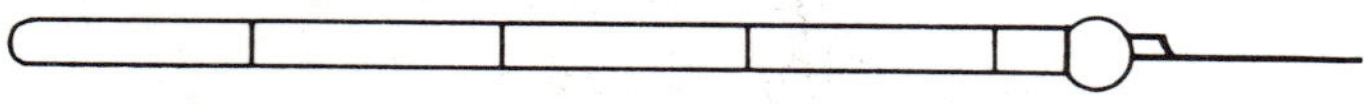

(5) a probe with several rods projecting from the end, like a fish spear;

(6) a short collapsible probe for easy handling;

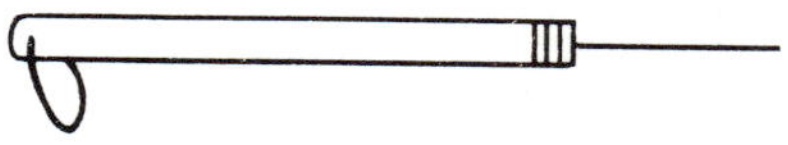

(7) a probe with prongs at both ends;

(8) the Vladimirov probe—equipped with a microphone for detecting delayed action fuses on mines and bombs.

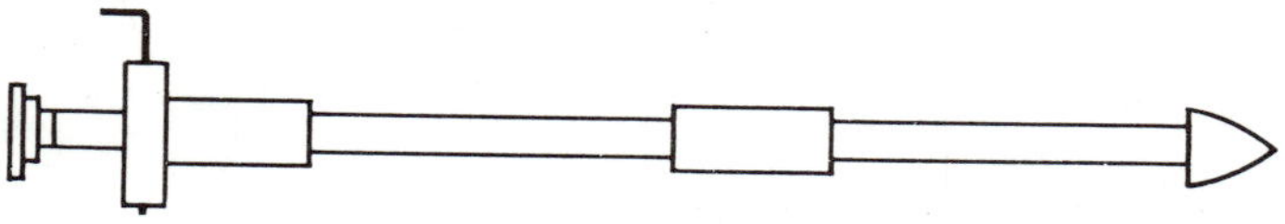

These probes, although somewhat hazardous to use, can be used to discover all types of mines, whether metallic or not.

In service with the Soviet and Warsaw Pact forces.

Optical Devices

These are to many the most fascinating of the combat aids. The apparent ease with which image intensification equipment can enable the soldier to see in the dark is remarkable to the layman. One unfortunate consequence, however, is that the soldier is no longer afforded the welcome protection that night used to bring. The parameters have changed—no longer is it the pitting of the soldier's wits against those of his enemy when, for example, he is carrying out a reconnaissance patrol. On patrol he cannot see properly, he cannot make best use of cover as he can by daylight and his enemy can probably watch his every move.

The great advantage of image intensification to the defender, or fire-controller, is that it is a passive system, hence the considerable encouragement given to its development, probably at the expense, rightly, of the infra-red devices which are active and thus detectable and limited in scope.

NSP2 infra-red night sight

USSR

Manufacturer State factories

The NSP2 is a conventional infra-red searchlight and optical sight which is used by most Warsaw Pact countries and mounted on AK-47 and AKM-type assault rifles and on light machine guns. It has also been observed being used in place of the normal optical sight on the RPG7V anti-tank grenade launcher and on the old RPG2. It is a rather cumbersome fitting with the 15 cm lens searchlight set above the 25·4 cm long optical sight, which is held 5 cm above the breech-cover of the rifle by the transformer container clamped to the left side of the breech, and an external power-supply cable. Although the NSP2 can be used on the obsolescent Nagant M1891/30 sniper rifle, it is not necessary for the SVD sniper rifle, which already has an infra-red element built into its telescopic sight, to which an infra-red lamp is added when required.

In service with the Soviet and other Warsaw Pact forces.

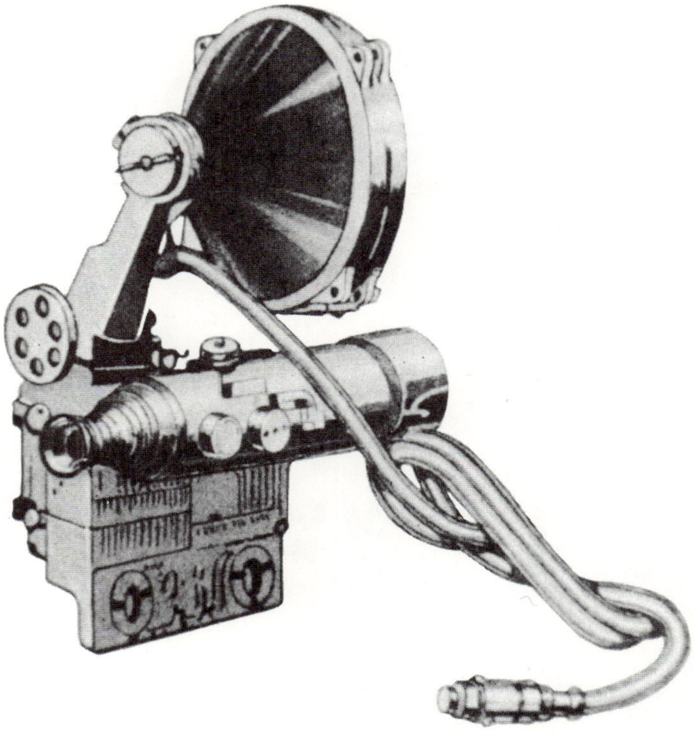

NSP2 infra-red night sight

PPN2 infra-red night sight

USSR

Manufacturer State factories

The Soviet PPN2 night sight, built on similar design principles to the NSP2, is a larger conventional infra-red device with more range. It consists, too, of a cumbersome assembly of a comparatively large infra-red searchlight, set above the long optical sight, which is held above the weapon breech-cover by the transformer-box clamped to the side of the weapon breech, and the external power supply cable. The PPN2 is designed for use on all medium and heavy machine guns.

In service with the Soviet and other Warsaw Pact forces.

PPN2 infra-red night sight

Pyrotechnics

There is a multiplicity of pyrotechnics varying from hand-held signal flares to coloured smoke bombs, from trip flares to illuminating missiles. Many are designed for civilian as well as for military use and the majority are efficient, reliable and most effective.

Despite the development of radar, seismic intruder devices and night vision systems, the soldier still needs to see for himself in the way he is used to seeing—with his eyes in daylight. Failing that, artificial daylight from a pyrotechnic is better than an eerie greenish image or a blip on a screen. The day this ceases to be true, pyrotechnics may cease to exist, but that day is a long way off.

Hand flare signal

USSR

Calibre	40 mm
Manufacturer	State factories

The 40 mm hand-launched flare signal is the bigger of the two devices of similar function (the other being a 30 mm flare) which have been taken into service recently in most Warsaw Pact countries, replacing the conventional signal pistol. The device consists of a two-part tube, one part being the disposable hand-held launcher, and the other the flare, made up of the illuminating charge, the propellant charge and a parachute to ensure that the best advantage is gained from the 10 sec duration of the illumination. A height of 200 m can be attained by the flare and a large number of different types of flare-filling can be obtained. Most other Warsaw Pact countries are beginning to produce these flares for themselves in the same design and sizes.

In service with the Soviet forces.

Signal pistol

USSR

Calibre	26·5 mm
Length	22 cm
Weight	0·8 kg
Manufacturer	State factories

This signal pistol is one of several similar equipments used in the Warsaw Pact countries, all being single shot, smoothbore weapons of the same calibre. The Soviet pistol, made of blued steel with a plastic grip, has no sights or trigger-guard and the trigger recesses into the front of the pistol-grip. To load, the breech is opened by pressing a stud located just below the trigger while downward pressure is applied to the barrel—the flare is inserted into the rear of the breech, which is then snapped shut. No safety-catch is provided. Many different types of flare can be fired from this pistol, but it is believed that in the Warsaw Pact forces signal pistols are gradually losing favour and are being replaced by the disposable and less cumbersome hand-launched flares.

The pistol is carried in a holster and a lanyard is normally attached to the butt-ring.

In service with the Soviet and some Warsaw Pact forces.

Infantry Support Vehicles

From Mark IX to IFV

Soon after the tank had shown its ability to dominate the battlefield of the First World War, it was also realised that, though it could take ground, it could not hold it. Infantrymen would still be required for that, and to do so, they must not only be able to keep up with the tanks, slow as they were, but also survive during the approach to their objectives.

Accordingly, in 1918 the British developed the Mark IX load carrying tank which was intended to transport the equivalent of an infantry platoon. However, the Mark IXs never saw service and were scrapped soon after the war.

Between the world wars, proponents of the armoured or tank division realised that its infantry component must be transported in vehicles. Various experiments were made, but cost was always a key factor. At the outbreak of the Second World War, the motor battalion of the British armoured division was carried in 'soft-skin' trucks while the panzer grenadiers of the equivalent German division were intended to be carried in armoured half-tracks. However, the half-tracks remained in short supply throughout the war and a panzer division was lucky to have more than a third of its establishment of vehicles. The Russians, with their apparent disregard for the lives of their men, carried their infantry into battle on the backs of the tanks. It was the United States Army, with the resources of American mass production, which was able to produce hundreds of half-tracks for its armoured divisions. Even in the half-tracks, the infantry was vulnerable to anything more than small arms fire and it was the Canadian First Army which, near Caen, used 'defrocked' Priest M3 self-propelled guns and gave the supporting infantry its first real protective vehicle.

The first attack using these vehicles was so successful and the casualties so minimal that it was decided to form two armoured personnel carrier regiments, one British and one Canadian. The vehicle chosen was the Ram, a Canadian version of the American M4 tank which suffered from being under-gunned. This tank, with its turret removed was called a Kangaroo. It offered the same protection to its infantry as the omnipresent Sherman—except from airbursts. Tactics varied from unit to unit and from action to action. The 49th Royal Tank Regiment in particular would motor its infantry as near to its objective as possible giving supporting fire from the machine guns with which the crews festooned their Kangaroos.

After the end of the Second World War, the role of infantry within the armoured division was much discussed but, among the NATO allies, no common doctrine was forthcoming in a situation that, sadly, exists still today. The Russians, being prodigal in their use of manpower, were less interested in preventing casualties and for some time were content that the infantry should continue to ride into battle on the tanks. But it was the threat of nuclear war that brought about a decision on both sides of the curtain. This was given impetus by the results of the test carried out in Nevada in 1955 using M75 and M59 APCs which showed that troops carried in such vehicles had an enhanced survivability in a nuclear battle. The M75, based on the Bulldog tank, had been the first APC to enter production in the United States following the outbreak of the Korean War. It was supplemented by the M59, but neither of these vehicles was entirely satisfactory, being underpowered and having high silhouettes; they were also expensive. It was the introduction into service of the M113 (which originally cost less than $25,000) that made it possible to produce large numbers of useful APCs. Although modified in many ways during the war in Vietnam, in a European context they are basically a battle taxi and their infantry will be expected to fight on foot. The use of aluminium armour has resulted in a lighter vehicle which is able to swim without flotation gear. Such vehicles are in service in many countries; they are still in production in their latest variants and seem likely to be around in the 21st century.

United States of America

In United States' service, the M113 is due to be replaced by a mechanised infantry combat vehicle (MICV) which has been around in various guises since the early sixties and the story of its development and the many changes that have beset the programme show the uncertainty that is felt over the use of mechanised infantry. The original problem of capacity has been resolved and the vehicle is expected to carry a section or squad, which is the load generally accepted by both NATO and the Warsaw Pact countries. The ability of the section to use its personal weapons is accepted also and all new infantry vehicles have fire ports. In addition, the American vehicle will mount a permanent weapon, such as the short M16 or the M231, though not many countries would wish to go to the expense of providing men with two weapons.

The new United States vehicle—at present known as the IFV (Infantry Fighting Vehicle) which is certainly an improvement on the last acronym of MICV (TBAT–2)—is now for reasons of cost going to supplement the M113. A new version of the latter had been announced called the IICV (Improved Infantry Combat Vehicle) which is to be fitted with a heavy weapons turret with either a cannon (20 or 25 mm) or a 40 mm automatic grenade launcher, plus a remotely controlled rear-mounted MG. The latter is likely to be either a 5·56 mm gun or 7·62 mm (NATO) FN MAG. By the time this appears in print some of these uncertainties will, happily, have been resolved. It is therefore likely that by the 1980s the US Army's mechanised infantry platoon will have four vehicles of which two will be M113 Als, one will be a M113 (IICV) and one a IFV; the IFV having a two-man turret with Tow, a Bushmaster cannon and a coaxial MG. The platoon commander's vehicle will also carry Dragon (ATGW) and Viper (LAW) plus possibly a sustained fire MG. The IFV turret must not be confused with the Improved Tow turret which is being fitted to M113s of armoured cavalry regiments. The number of anti-tank weapons that a US mechanised infantry battalion will be able to field is most impressive. Yet, there are also likely to be two Bushmaster (type) cannons in each platoon to deal with enemy APCs.

USSR

The first post-war Russian APC, the BTR-152, was a wheeled vehicle, but it was followed by the BTR-50 which is tracked. Initially, neither of these vehicles had overhead armour and only some of the latest model BTR-50s have NBC protection. The next Russian APC, the BTR-60 is again wheeled and intended for service with Soviet motor rifle regiments. The belief of Glavnyy Voyennyy Sovyet (the main Soviet military council) that the Soviet forces could win a nuclear war has resulted in the BMP design which is intended to exploit the results of the use of tactical nuclear weapons. However, the swing back to the concept of a conventional battle and the lessons of use of the BMP in the 1973 war have raised doubts among Soviet planners regarding the BMP. At present the motor rifle division normally seems to have a BMP regiment and a BTR-60 regiment. This must cause some problems for the divisional commander as the range of the latter is 500 km to the former's 300 km and, inevitably, there are other differences in performance. Soviet military writers have been considering how best to employ the BMP. It appears that it was originally thought possible to use BMPs without main battle tank support, but it now seems the maximum number that should manoeuvre in this way is a battalion of some 31 vehicles. There have been reports, too, of concern over the danger of fire when launching Sagger near the magnesium armour.

It is true that Soviet commanders are averse to using dismounted infantry in an advance as they consider it vital to maintain the momentum of the attack at all times and men on foot cannot do that. It is certain, too, that companies or battalions of BMPs will be sent to the flanks, ahead of the main advance, to live on their own. Destruction of the BMPs, thus causing infantry to dismount, could be most effective in disorganising a Soviet-type attack.

One basic lesson the Soviet Army has appreciated from the 1973 war is the need for all arms to operate together and it is reported that a BMP company has recently been added to each tank battalion in a tank division.

Information about new developments in Soviet designs for future infantry combat vehicles are, as always, scanty, but there are reports of the M-1970 gun-tractor appearing in the motor rifle battalion and that some marks of the BMP are appearing fitted with a cannon. It is also likely that it will not be long before the Sagger is replaced by a newer ATGW. It will be particularly interesting then to note whether the armour will be improved on any BMP replacement.

France

Since the Second World War, France has been the third largest producer of APCs and her products can be found worldwide. Initially, wheeled vehicles were produced, such as the EBR Sahara for colonial policing, but military policy later decided that a tracked vehicle was needed to operate forward with tanks in the panzer grenadier role. Hotchkiss developed a reconnaissance vehicle for the French Army, but it was not adopted. The first design to go into production for the French Army was the AMX-VCI which was based on the AMX-13 light tank chassis. It has been described as the first mechanised infantry combat vehicle. The infantry are seated down the centre back to back, each man being opposite a firing port. Shortly after the first models were produced, they were given turrets.

Since then, France has continued to produce tracked infantry combat vehicles, but cost has prohibited all mechanised infantry units being equipped with the latest vehicle, the AMX-10, and it has been decided to go for a mix of tracked and wheeled vehicles. Operationally, it is intended that AMX-10 will be deployed forward with the main battle tanks while the wheeled VAB will be an armoured 'taxi' and logistic support vehicle.

Germany (FGR)

The West German Army was quick to seize on the MICV, like the Soviet Army perhaps, drawing on shared experiences on the Eastern Front in World War II. After many trials with the Swiss firm of MOWAG, they decided on the Schutzenpanzer Neu, now better known as the Marder. A panzer grenadier battalion has over fifty Marders at present, though 'reconstruction' may reduce this slightly. It is an extremely expensive vehicle and provides its ten man crew with a variety of weapons including a 20 mm cannon and a rear MG, both remotely controlled from within the vehicle. More recently, vehicles have been seen with a Milan launcher. The West German Army intends its Panzergrenadiers to hold ground dismounted but, in a counter-attack role, they will be able to fight from behind their armour.

United Kingdom

The United Kingdom has also decided that it is necessary for an APC to be tracked if only to have a similar cross-country performance to the tanks it will accompany. Saracen and the Humber 'Pig' are both basically internal security vehicles, though the former was attractive to small armies in its APC role. The result, therefore, was the production of the FV432 and family. Like the M113, it is a battle-taxi with no gun ports and, though various turrets have been fitted including a 30 mm Rarden, the resulting loss of space has resulted in either a small GPMG turret or a pintle mounting. The UK Army is shortly to announce its decision on a new MICV but, because of defence cuts, production may be deferred for some time yet. It is expected that it will have to be a larger vehicle than the FV432, with an anti-APC main armament. The FV432 is of all steel construction and needs flotation screens, but it is possible that, following experience with the Scorpion family, the United Kingdom will choose aluminium armour.

China

The Chinese People's Liberation Army is known to have Soviet type vehicles, based on the BTR-40 and the BTR-152 from before the Sino-Soviet split and a national design, the K-63, has been in service for some years. There have been reports of a newer vehicle but, as is usual with CPLA equipment, information is scanty. It seems very likely that the CPLA would regard its vehicles more as APCs rather than as IFVs.

Other countries

Another recipient of Soviet vehicles, India, is believed to be in the process of producing its own vehicle. In this case, the assistance is now thought to have come from Japan. Japan itself has developed two tracked APCs and the latest, the Type 73, has firing ports, unlike its predecessor the SU-60. The newer vehicle has aluminium armour while the SU-60 was of steel construction.

One of the first military vehicles a developing nation builds is an APC and Brazil, which is attempting to build up an armaments base in South America, has an extremely interesting six-wheel APC known as the Urutu EE-11. Egypt also has a home-designed vehicle the Walid, but this is no longer in production. While in NATO, the Netherlands has manufactured the YP-408, Italy makes its own version of the M113 and has a MOWAG variant, the Tifone, and a Fiat wheeled APC on trials. This independent line has also been followed by Warsaw Pact countries. The Czechoslovakian range of OT (APC) vehicles is considered to be an improvement on their Soviet originals and has been purchased by other Warsaw Pact countries.

Among the industrial non-aligned nations, mention has already been made of the Swiss firm of MOWAG, which has a range of wheeled and tracked vehicles to its credit. Canada has just announced its intention to build a version of the Piranha under licence. Other countries are believed to have similar arrangements. Sweden naturally has its own vehicle, the Pbv 302, as has Yugoslavia in the M60. The former is very much a homegrown variety, while the latter is based on Soviet technology. The newer Yugoslavia M-980 is thought to be an amalgam of Russian and French technology. Perhaps one of the most interesting new arrivals on the scene is Eire with its Timoney, which is basically an IS vehicle. It is being manufactured in one variant by the Belgian firm of Beherman Demeon. Another country, Austria, has also manufactured its own APC, but it is doubtful whether this small scale production can continue in the future.

Looking to the future of the armoured infantry carrying vehicle, it would seem that a large scale production of sophisticated vehicles (like the Marder) will not happen again. It would also seem that, when the next generation British vehicle is announced, we may get an idea of current NATO thinking. The United States Army obviously sees a tank-killing role for the IFV with its combined anti-tank Tow and Bushmaster cannon, but it has taken pains to emphasize that it is the latter weapon that is the main armament.

So, just as the tank has proved to be the main tank killer, so it seems that the IFV is going to be the primary anti-infantry fighting vehicle weapon. The men of the mechanised infantry battalions will have a series of sophisticated weapons to use for their many tasks which will include defending tanks at night, holding vital ground, tank hunting, providing local patrols and exploiting success. The last being, perhaps, the primary role of the better Warsaw Pact motor battalions. For NATO and Warsaw Pact countries, the common need to survive, whether on a conventional battlefield or in NBC conditions, or both, will dictate their next vehicles.

For other countries, the wheeled APC, like the new South African Ratel (which can be used both for war and counter-terrorist activities) is likely to be the most effective. Even if the gun tank is superseded by a missile carrier, men will, in the not-too-distant future, continue to be needed to take and hold ground and good man-carrying vehicles are going to be needed all the more. The armoured infantry carrying vehicle is therefore likely to outlive the tank which brought it into existence.

On Wheels to War

Many readers will have heard of the 'taxis to the Marne', the motley collection of vehicles that helped General Joffre turn the tide against the advancing German armies in 1914. Pictures of open-top London buses carrying British troops to the front line illustrate how infantry have been carried to battle since the coming of the combustion engine. Between the wars there was a policy for infantry battalions to have their own transport section to carry the platoons, companies and the battalions' basic supplies and heavier items of equipment. Within NATO, with its emphasis on the mechanised battalion, the centralising of transport has been taking place under such service units as the Royal Corps of Transport (in the UK) in attempts to cut down the fighting units' tails and to achieve greater cost effectiveness.

Exactly how vital supplies will be brought forward is still under study, but the result is that the few infantry battalions that armies now retain on stations no longer control the 15 cwts, 3 tonners, or command and weapon carriers that were part of the scene at so many barracks in the past. The one vehicle that is still to be seen as frequently as it was in the past is the light utility vehicle or, more colloquially, 'the Jeep'. Because of the tendency, now, to move to larger vehicles, as well as to try and remove them from direct charge, today's infantry will be carried in 4-ton Bedfords, 10-ton AEC Militants, 6-ton Berliet, 4-ton Saviem, $2\frac{1}{2}$-ton M-34s, 5-ton M-54s, and so on, but they are being withdrawn from the infantry unit establishment.

One impressive fact about the latest generation of cargo carrying trucks is their cross-country performance. Most of these vehicles would be able to supply the forward troops in conditions that would bog down and halt their fore-runners. Multi-wheel drive is universal and the power-to-weight ratio is greatly improved, but it is certain that no army is going to bother to produce a simple 'troop carrier' when, in a war situation, it would be possible to call on hundreds of civilian buses and coaches to move infantry, if it was necessary or desirable.

With the cost of modern weapons rising so fast and defence budgets being cut, army planners see a waste of money in having vehicles that will not be fully utilised in peace time—and possibly in war. There is however a need for that 'Jeep'—a small adequately powered vehicle that can carry radios, carry controlling staff around the exercise ground, police the column on the move, patrol and carry out all the other operational tasks that need wheels to save time and aid efficiency. It must also be helicopter-portable.

The first country that produced its own 'Jeep' following the end of Lend-Lease was the United Kingdom and its Land Rover now serves in nearly one hundred and fifty countries. There are several versions, $\frac{1}{4}$-ton, $\frac{1}{2}$- and $\frac{3}{4}$-ton with a short and long wheel base, $\frac{1}{2}$ tonne and for non-infantry duties the 1 tonne. France produced its own Jeep, the Hotchkiss M201, Italy, the $\frac{1}{2}$-ton Campagnola and West Germany, the Auto Union 'Munga'. Behind the Iron Curtain Jeeps were replaced with the highly efficient UAZ-69, the larger GAZ-69 and now the new UAZ-469. East Germany has produced the P2M and P3, while in Japan Mitsubishi produced a $\frac{1}{4}$-ton utility and Nissan a $\frac{1}{2}$-ton vehicle. The latter is made in India and offered for sale overseas. Many countries have their own motor industries and there have been several other home-grown products. Some, like the Austrian Steyr-Puch Haflinger, have remarkable performances.

Concern has been shown in some quarters that commanders travel the main battle area in unarmoured vehicles. The United Kingdom, for example, produced an armoured command/scout vehicle in the Vixen which was a victim of defence cuts, though it still uses the successful Ferret. However, with modern communications, such as television links which bring him once more into (electronic) personal contact, the tendency is less for commanders to go and see. For peacetime needs the Jeep/Land Rover should provide a satisfactory answer for many years.

Oddly enough, one vehicle that may return in the coming years is the motor-cycle. The US Army is particularly interested in its possibilities and the new IFV/Scout has been designed to carry one. In planning for a European war, soft skinned vehicles must be assumed to be vulnerable. West Germany has a prototype Transportpanzer, but the cost makes the likelihood of many entering service slight. Airborne troops have specialised load carriers like the West German Kraka, the US Army's mechanical mule, there are the private venture Saboteur and Argocat; for special terrain Canada has XM-571 Dynatrac, Finland the Valmet 865 and Sweden the Volvo Bv 202.

It seems likely that Western armies, at least, will see a reduction in their integral soft-skinned infantry vehicles to a minimum establishment. Peace time moves will be carried out by civilian vehicles under contract and in a 'three week' war, all the NATO infantry of such groups as Britain's Field Forces will be able to do is sit tight and hold ground as a pivot, or datum point, for the armoured and air battles.

OT-62

Czechoslovakia

Type	Tracked armoured personnel carrier
Crew	Commander, driver and eighteen (OT-62A and B) Commander, driver, gunner and twelve (OT-62C)
Weight	15,000 kg (OT-62B)
Length	7 m
Width	3·23 m
Height	2·23 m (OT-62B), 2·73 m (OT-62C)
Engine	300 hp six cylinder model PV-6
Maximum speed	60 kph
Range	500 km approx
Armament	OT-62B small turret with 7·62 mm M59T MG and 82 mm RCL OT-62C turret with 7·62 mm PKT and 14·5 mm KPVT HMG
Armour	14 mm maximum
Manufacturer	State factories

The OT-62, which is based on the Soviet BTR-50PK, has been a very successful vehicle—unlike the BMP-1, OT-62s captured by the Israelis have been taken into service. It entered production in Czechoslovakia in 1963 and has been adopted by that country and Poland in place of the 50PK. It has an all-welded hull and is fully amphibious. Its more powerful supercharged engine gives it a better cross-country performance than the 50PK and the Polish army has adopted the WPT-TOPAS for the recovery role.

The 62A has no fixed armament, but the T-21 82 mm recoilless rifle can be fired from it. Some production batches do not have the side entrance doors. The driver is centrally placed with the gunner to his right and the commander on his left. The recoilless rifle of the 62B is fixed to the right, outside the small turret and so, to reload, the gunner must expose himself. There are two hatches behind these positions which allow six infantry to use their weapons over the side of the vehicle. The 62C has a centrally placed turret behind the commander and its 14·5 mm machine gun has a maximum elevation of 78 degrees for use in an anti-aircraft role.

In service with the Bulgarian, Czechoslovak, Egyptian, Hungarian, Indian, Iraqi, Israeli, Libyan, Ethiopian, Polish and Romanian forces.

OT-62 with the T-21 82 mm RCL

OT-64 SKOT

Czechoslovakia

Type	Wheeled armoured personnel carrier
Crew	Commander, driver and eighteen
Weight	14,300 kg (OT-64A)
Length	7·44 m
Width	2·5 m
Height	2·03 m (OT-64A)
Engine	180 hp Tatra T 928-14 diesel
Maximum speed	93 kph
Range	700 km
Armament	One 7·62 mm MG (OT-64A)
Armour	10 mm maximum
Manufacturer	State factories

The OT-64 is the result of a joint design effort by Czechoslovakia and Poland and is used in their armies in place of the BTR-60 to which it bears a quite close resemblance. It has a complete NBC system and has been exported to friendly countries and is sometimes in service alongside the BTR-60, its Soviet counterpart.

The OT-64 Model 3 has a turret with a KPVT 14·5 mm HMG which reduces the number of infantry carried from 18 to 15 (the KPVT is effective against most APCs). The OT-64 Model 4 of the Polish army has a different turret for the KPVT with a higher elevation. Some Model 4s have been observed with a Sagger ATGW on either side of this turret. A number of earlier Model 1s have been converted to an anti-tank role with twin Sagger launchers at the rear and several reloads carried internally.

In service with the Czechoslovak, Egyptian, Ethiopian, Hungarian, Indian, Libyan, Polish, Syrian and Ugandan forces.

OT-64 SKOT 2-AP with the 14·5 mm MG

BTR-152

USSR

Type	Wheeled armoured personnel carrier
Crew	Commander, driver and fifteen
Weight	8,950 kg
Length	6·83 m
Width	2·23 m
Height	2·05 m
Engine	110 hp Zil petrol engine
Maximum speed	75 kph
Range	650 km
Armament	MG mounting
Armour	13·5 mm maximum
Manufacturer	State factories

Looking very like a wheeled version of the famous American half-track of the second world war, the BTR-152 appeared not long afterwards and, probably, in greater numbers. It has a hull of all-welded steel construction with the driver and commander sitting side by side at the front, each with a side door. The infantry sit in an open-top compartment behind them. There are twin rear entrance doors. It has been used as a command vehicle as well as with HMGs for anti-aircraft defence and fire support. Various numbers of rifle calibre machine gun fittings have been used in local conditions. One version, the 152K with an armoured roof, was produced, but not it seems in great numbers.

In service with the Afghan, Algerian, Bulgarian, Cambodian, Ceylonese, Congolese, Cuban, Cypriot, DDR, Egyptian, Ethiopian, Guinean, Indian, Irani, Iraqi, Israeli, Lebanese, Mongolian, North Korean, North Yemeni, Polish, Romanian, Somalian, Soviet, Sudanese, Syrian, Tanzanian, Ugandan, Yugoslav forces and, as the Type 56, with the Chinese and Albanian forces.

BTR-152

BTR-50

USSR

Type	Tracked armoured personnel carrier
Crew	Commander, driver and fourteen
Weight	14,200 kg
Length	7·08 m
Width	3·14 m
Height	1·97 m
Engine	240 hp V-6 diesel
Maximum speed	45 kph
Range	250 km
Armament	Pintle mounted MG
Armour	14 mm (maximum)
Manufacturer	State factories

An adaptation of the PT-76 light tank to the APC role, this vehicle suffers some shortcomings as a result. It has an all-welded steel hull and the crew compartment is at the front with the engine and transmission at the rear. In the Warsaw Pact forces, the BTR-50 has an armoured roof with two hatches (BTR-50PK) and can operate under NBC conditions. Some of the open top early versions, 50P and PA, may still be in service but it is doubtful whether they are in service with Soviet forces in Germany.

There are six road wheels and the driving sprocket is at the rear; the track has no return rollers. BTR-50s are amphibious and are capable of 11 kph in water by means of twin jets of the same type as those used on the PT-76. Reports from the Arab-Israeli wars indicated that the former used these vehicles very much as battle-taxis, with as many as twenty troops aboard. In Israeli service, the roofs are removed and several MGs are fitted, but the crew is kept to a maximum of two plus ten infantry.

The BTR-50PU command version can be found with many different units of Soviet trained forces as the vehicle seems well adapted to this role. Versions have been noted with five aerials and they are known to carry a complete navigation system with course plotter and co-ordinate indicator.

On service with the Afghan, Algerian, Finnish, GDR, Irani, Israeli, Polish, Romanian, Somalian, Sudanese, Syrian, Ugandan, Vietnamese, Soviet, Yugoslav and, possibly, the Chinese forces.

BTR-50P (with open top)

BTR-60

USSR

Type	Wheeled armoured personnel carrier
Crew	Commander, driver and fourteen
Weight	10,000 kg (with armoured roof only)
Length	7·56 m
Width	2·82 m
Height	2 m (2·31 m with turret)
Engine	Two 90 hp GAZ 49B
Maximum speed	80 kph
Range	500 km

Armament	BTP-60P—various pintle mounted MGs BTR-60PK—12·7 mm DShK MG or 7·62 mm pedestal mounted plus pintle mounts BTR-60PB—turret with 14·5 mm KPV HMG and 7·62 mm MG
Armour	10 mm (14 mm on turret)
Manufacturer	State factories

This is probably the most widely used APC supplied by the USSR, together with its Czechoslovak version, the OT-64. It first entered service with the Soviet Army in 1960.

It has a hull of all-welded steel with the driver and commander sitting at the front of the vehicle—the commander on the right, behind vision blocks. The section in a Warsaw Pact motor battalion is normally nine men of which one is a vehicle driver, but the exact complement in a BTR-60 is uncertain. The two engines which power all eight wheels are at the rear, while the front two pairs are power-steered.

There are three firing ports at each side of the hull and roof hatches. There are variations in these from production batch to batch as in many other vehicles. Entrance and exit are normally over the side, but the 60PB has two small doors near the turret. All models are fully amphibious and propulsion is by water jet. Like all later Soviet AFVs, the PK and PB are fitted for NBC warfare when supplied to Warsaw Pact forces. Also like all modern Soviet wheeled vehicles, the tyre-pressures can be controlled by the driver. The BTR-60 is supposed to be able to operate on one engine, but this would have to be under good road conditions.

BTR-60PK

Production of the BTR-60 is believed to have ceased in the Soviet Union, but it is likely to remain in service for many years. Its replacement in the motor battalion is likely to be a mix of BMP and M-1970.

In service in Afghan, Algerian, Bulgarian, Cuban, DDR, Egyptian, Hungarian, Irani, Iraqi, Israeli, Libyan, Mongolian, North Korean, Romanian, Somalian, Soviet, Syrian, Vietnamese and Yugoslav forces.

BTR-60P (early open top version)

BMP-1 and BMP-2

USSR

Type	Tracked mechanised infantry combat vehicle
Crew	Commander, driver, gunner and eight
Weight	12,500 kg loaded
Length	6·75 m (BMP-1)
Width	3 m
Height	2 m
Engine	280 hp model V-6 diesel
Maximum speed	55 kph
Range	300 km
Armament	One 73 mm low pressure smooth bore One 7·62 mm PKT MG Launcher rail for ATGW (Sagger) above the 73 mm barrel
Armour	14 mm maximum (magnesium)
Manufacturer	State factories

The appearance in large numbers of the BMP (M-1967 APC) indicates a new approach to infantry mobility in the Soviet Army—that of motoring on to an objective with tanks, using prophylactic fire from their own and the infantry's weapons.

Design and construction of the BMP was a new departure for the Soviet State factories, assisted (it is said) by Fiat mass production techniques. The MICV has a hull of magnesium armour. There are six road wheels, front-drive sprocket and rear idler. The engine is mounted at the front to the right of the driver, behind whom is the commander with the one-man turret with its 73 mm and coaxial MG on the centre line. Because the turret is set low and far back, there must be minimal depression available to the low pressure gun, a feature which it has in common with Soviet tanks. The main compartment has seating for eight men back to back. Some reports from Syria have suggested ten being carried, but this is very unlikely. The front men sit by the MG side ports, either 7·62 mm PKs or RPKs, which are remotely controlled from periscopic sights. The remaining six men have ports for their Kalashnikov assault rifles and there is a further rifle port in the left-hand rear door.

There is a central fume-extracting system. There are four roof hatches; on some vehicles the hatch cover of the right hand rear man appears to be altered so he can use a SAM launcher with greater ease and safety; each vehicle carries one of these or a RPG-7.

Many correspondents after the 1973 war reported on the BMP as the Soviet wonder weapon, but later assessments suggest that it is less than ideal. The magnesium armour can burn—and did—the fuel tanks in the rear doors are an added fire hazard and when closed down visibility is restricted and quarters are very cramped. The BMP is amphibious, propulsion is by the tracks, but it floats so low in the water that a snorkle is needed and there were reports of vehicles being swamped. The Israeli Army has resisted taking any captured BMPs into service.

BMP-1

The BMP-2 has a lengthened front and a longer rear deflector shroud—both improvements are intended to enhance its waterborne performance. There are also reports of a new main gun and a replacement missile to the present Sagger. BMPs in service with Warsaw Pact countries have been reported as fitted with a 23 mm (?) cannon in place of the 73 mm. This may be in part because the range of the larger gun is only some 900 m.

In service with the Czechoslovak, Egyptian, GDR, Iraqi, Libyan, Polish, Soviet, Syrian and Ugandan forces.

BMP-2 (fitted with Sagger)

M-1970 (MT-LB)

USSR

Type	Tracked armoured personnel carrier
Crew	Commander, driver, gunner and ten
Weight	12,000 kg
Length	6·35 m
Width	2·8 m
Height	2·28 m
Engine	200 hp IZ diesel
Maximum speed	60 kph
Range	450 km
Armament	Turret mounted 7·62 mm
Armour	Unknown

(All details above are provisional.)

This vehicle was first seen in use with artillery battalions. It is of all-welded construction, with the driver seated forward on the left-hand side. The turret is offset to the right. There are twin doors to the troop compartment and roof hatches. It has firing ports down the side which indicate that it may be used with motor battalions. It is possible that production will be increased, especially if the stories of dissatisfaction with the BMP are true. It is fully amphibious and uses its tracks for propulsion. It has NBC protection.

In service with the Warsaw Pact forces.

M1970 (MT-LB)

M60

Yugoslavia

Type	Tracked armoured personnel carrier
Crew	Commander, driver, gunner and ten
Weight	9500 kg
Length	5·05 m
Width	2·75 m
Height	1·8 m to compartment roof
Engine	140 hp Famos diesel
Maximum speed	45 kph
Range	400 km
Armament	One 12·7 mm MG. A bow MG may be fitted
Armour	25 mm maximum
Manufacturer	State factories

(All above details are provisional)

M-60 with the 12·7 mm MG

Though designed and produced in Yugoslavia, the M60 (previously known as M-1965 and M-590) uses a Soviet track system taken from the SU-76 self-propelled gun. The M60 is said to be fully amphibious and is driven by its tracks. The driver sits at the front on the left. The commander has a hatch behind the driver and on his right is the HMG which is capable of high elevation against aircraft. There are three firing ports down each side of the hull and one in each of the rear doors. The diesel engine is of the Austrian Saurer design. It is not thought that many of these vehicles have been made.

In service with the Greek Cypriot Defence and Yugoslav forces.

M-980

Yugoslavia

Type	Tracked mechanised infantry combat vehicle
Crew	Commander, driver, gunner and eight
Weight	12,000 kg
Length	6·5 m
Width	2·9 m
Height	2·2 m to top of turret
Engine	275 hp HS 115-2 diesel
Maximum speed	70 kph
Range	500 km
Armament	One 20 mm cannon, one 7·62 mm coaxial and two Sagger ATGW
Armour	Not known
Manufacturer	State factories

(All the above details are provisional)

M-980 with two Sagger ATGW

First seen on a May Day parade in 1975, the M-980 is believed to have now entered service in some numbers. Much of the construction is probably based on the French AMX-10P (qv).

The driver is at the front left with the commander behind him. The turret is in the centre of the hull and behind the commander to his right. Photographs have shown vision/gun ports in the side and rear of the hull. Entrance to the troop compartment is from the rear and the vehicle is thought to be fully amphibious and to have a NBC system.

In service with the Yugoslav forces.

Glossary and Abbreviations

AA	Anti-aircraft
ACP	Automatic Colt pistol
AP	Anti-personnel
AP	Armour piercing
APC	Armoured personnel carrier
APDS	Armour piercing discarding sabot
APV	Armoured personnel vehicle
AT	Anti-tank
ATGW	Anti-tank guided weapon
AV	Anti-vehicle
Ball ammunition	Live small arms ammunition
Ball cartridge	The standard anti-personnel round
Ballistite round	Bulletless cartridge used for firing rifle grenades (Blank)
Benelux	Belgium, Netherlands and Luxembourg
Blank cartridge	Cartridge used to simulate live ammunition or to launch grenades
Blowback	Type of weapon operation that uses the gas to operate the cycle without any form of locking system
BT	Bullet trap (see grenades section)
Bulleted blank	Type of blank which has a 'safe' form of bullet so that an automatic weapon will function
Bull pup	Type of weapon that has the normal position of the magazine and pistol grip/trigger reversed
C	Centigrade
Cal	Calibre
Carbine	Short version of a rifle
Centre fire	Standard form of metallic cartridge with the primer in the centre
cm	centimetre
Compensator	Muzzle device fitted to reduce recoil or muzzle climb
CP	Candle power
CVR (T)	Combat vehicle reconnaissance (tracked)
Cyclic rate	Rate at which an automatic weapon fires
DC	Direct current
DDR	Deutsche Demokratische Republik (East Germany)
Delayed blowback	Where the cartridge produces too high a pressure and thus a dangerous early opening of the breech. In a simple blowback, some form of delay is used
Effective range	Range at which the weapon may be used with reasonable results
Exterior ballistics	Science of ballistics applied to the projectile in flight when it has left the barrel
Extreme range	Range which the projectile will not exceed
Finabel	France, Italy, Netherlands, Belgium and Luxembourg
FGR	Federal German Republic
Flechette	Small diameter arrow-type of projectile that is usually fired with a discarding sabot
Flash hider	Device used to prevent (or reduce) the muzzle flash from a weapon
Fluted chamber	With the use of high-pressure cartridges in delayed-blowback weapons, some problems exist with initial extraction. A fluted chamber floats the case off the chamber walls by bleeding gas down flutings in them
GDR	German Democratic Republic
gm	gramme
GPMG	General purpose machine gun
Grip safety	An additional safety that requires the firer to hold the grip while operating the weapon
HE	High explosive
HEAP	High explosive anti-personnel

HEAT	High explosive anti-tank
HEPAT	High explosive penetrating anti-tank
HEP – T	High explosive penetrating—tracer
HESH	High explosive squash head
Hold open	Device to keep the breech open normally on the last shot from the magazine
hp	horse-power
I	Incendiary (APT-I)
IFF	Interrogative friend or foe?
IFV	Infantry fighting vehicle
IICV	Improved infantry combat vehicle
in	inch
IR	Infra-red
kg	kilogramme
Kurz	Short (German; used in designations)
Largo	Long (as of a bullet, 9 mm largo)
lb	pound
LMG	Light machine gun
Locked breech	Type of breech mechanism that is positively locked at time of firing
m	metre
MG	Machine gun
MICV	Mechanised infantry combat vehicle
mil	A measurement of angle
min	minute
Mk	Mark
mm	millimetre
MMG	Medium machine gun
MTBF	Mean time between failure
m/sec	metres per sec
Muzzle brake	A muzzle attachment that reduces recoil
Muzzle energy	Energy of the projectile as it exits the muzzle
Muzzle velocity (MV)	Velocity of the projectile as it exits the muzzle
NATO	North Atlantic Treaty Organisation
NBC	Nuclear, biological and chemical
Obturation	Sealing of the breech against the gas from the propellant
Parabellum	Designation applied mainly to the 9 mm pistol round
PEPA/LP	Projectile empené à propulsion additionelle/longue portée (finned with additional propulsion)
Primacord	Instantaneous fuse
Propellant	Substance that is burnt to provide the gas to project the projectile
PRPA	Projectile rayé à propulsion additionelle (rifled with additional propulsion)
RCL	Recoilless rifle
Recoil operation	Weapon system that uses the recoil energy to operate
rpm	rounds per minute
sec	second
Selective	Weapon that can be fired both single shot and fully automatically
Selector	Device on the weapon to enable the firer to choose the type of fire required
Self-loading	Any weapon that reloads without the firer having to do so manually
Shaped charge	Explosive charge with a cone-shaped cavity that intensifies its explosive effect
SL or SLR	Self loading or self loading rifle
SMC	Sub machine carbine
SMG	Sub machine gun
SMLE	Short magazine Lee Enfield (rifle)

T	Tracer
UK	United Kingdom
US/USA	United States/United States of America
USAAF	United States Army Air Force
USSR	Union of Soviet Socialist Republics
Velocity	Speed of an object, normally a projectile
v	volt
WP	Warsaw Pact
× 2/× 3	Magnification of two/of three (etc)

Index